Olive Oil Phenolics as Potential Therapeutical Agents

Olive Oil Phenolics as Potential Therapeutical Agents

J. M. Arbonés-Mainar,
M. A. Navarro,
J. M. Lou-Bonafonte,
M. V. Martínez-Gracia
and
J. Osada

Nova Science Publishers, Inc.
New York

LIBRARY OF CONGRESS CATALOGING-IN-PUBLICATION DATA

Olive oil phenolics as potential therapeutical agents / J.M. Arbonés-Mainar ... [et al.], (authors). p. ; cm.

Includes bibliographical references and index.

ISBN 978-1-60692-220-0 (softcover)

1. Phenols in the body. 2. Phenols--Therapeutic use. 3. Olive oil--Physiological effect. 4. Olive oil--Therapeutic use. I. Arbonés-Mainar, J. M.

[DNLM: 1. Plant Oils--therapeutic use. 2. Phenols--chemistry. 3. Phenols--therapeutic use. QU 86 O48 2009]

QP801.P4O45 2009 547'.632--dc22 2008038245

Published by Nova Science Publishers, Inc. ✝ New York

Contents

PREFACE

Our knowledge of the composition of phenolic compounds present in olive oil has been profoundly increased in recent years due to sophisticated analytical technologies. These compounds have been clearly shown in vitro to possess strong antioxidant and free-radical scavenging activity. Moreover, recent studies have evidenced that these compounds are absorbed, metabolized and transported in lipoproteins. A great deal of experimental work has evidenced potential roles for these compounds in protecting the cardiovascular system: 1) by direct antioxidant defence, especially against lipid peroxidation and DNA oxidative damage; 2) by preventing endothelial dysfunction, including increased nitric oxide (NO) production and inducible NO synthesis, and quenching vascular endothelium intracellular free radicals; 3) by inhibiting platelet-induced aggregation; 4) by enhancing the mRNA transcription of the enzymes involved in antioxidant defence; 5) by inducing apoptosis and anti-proliferative effects; 6) by presenting antimicrobial and antiviral activity and modulating immune system function. In animal models, emerging beneficial effects on the control of blood pressure have been observed, but conflicting results on the role of olive phenolics in the management of atherosclerosis are also presented. In conclusion, research has robustly shown that these natural chemical compounds may play important roles in vivo but much effort still remains to be done before a definitive recommendation regarding their use in clinical settings.

ABBREVIATIONS

COX-2	cyclooxygenase 2
DPPH, 2	2-diphenyl-1-picrylhydrazyl model

EVOO	extra virgin olive oil
HT	hydroxytyrosol
iNOS	inducible nitric oxide synthase
8-iso-$PGF_{2\alpha}$	8-iso prostaglandin $F_{2\alpha}$
LDL	low density lipoproteins
L-NAME	N-nitro-L-arginine methyl ester
NOAEL	non observed adverse effects level
OMW	olive mill waste waters
PUFA	polyunsaturated fatty acids
ROS	radical oxygen species
TBARS	thiobarbituric acid reactive substances
TG	triglycerides
VLDL	very low density lipoproteins

Chapter 1

INTRODUCTION

Cardiovascular diseases are the primary cause of death and morbidity in Europe, showing a clear West-East gradient with high rates in Eastern Europe. In spite of the decreasing trend in age-adjusted cardiovascular disease mortality in Western European countries, an increase in the number of cardiovascular patients is still expected in the future because of a burgeoning ageing population. Consequently, the health care and social costs of these diseases will increase dramatically [1].

In comparison to typical modern European diet, the traditional Mediterranean diet of Greece is based on a high intake of fruits, vegetables, nuts and cereals; more olive oil (40% of calories) and olives; less milk, but more cheese; more fish; less meat; and moderate amounts of wine. Analyses of this dietary pattern reveals a number of protective substances, such as selenium, glutathione, a balanced ratio of (n-6):(n-3) essential fatty acids, high amounts of fiber, antioxidants (especially resveratrol from wine, and biophenols from olive oil and other vegetables), and vitamins E and C, some of which have been shown to be associated with lower risk of cancer [2]. Epidemiological data has also shown that the Mediterranean diet has significant protective effects against coronary heart disease and other chronic diseases [3-5]. It is currently believed that the phenolic fraction (e.g flavonoids, simple phenols), along with the high intake of anti-cancer agents (e.g. squalene and terpenoids), as well as the peroxidation-resistant lipid oleic acid of olive oil, are mainly responsible for confering its health-promoting properties [6]. Indeed, it seems probable that olive and olive oil consumption in southern Europe represents an important contribution to the beneficial effects on health of the Mediterranean diet. In this regard, it has been demonstrated that high consumption of extra-virgin olive oils (EVOOs), which are particularly rich in these hydrocarbons, biophenols, tocopherols, sterols, triterpenoids and other components usually found in traces, should afford considerable protection against cancer, coronary heart disease and ageing, principally by inhibiting oxidative stress [7].

The present review will focus on the biological effects of phenolic compounds and their potential application as therapeutical agents. A recent review addressing the activity of triterpenoids and other unsaponifiable compounds can be found in [8].

Chapter 2

PHENOLIC COMPOUNDS IN OLIVE OIL, OLIVE OIL WASTE-WATER AND TABLE OLIVES

Knowledge of the variations in phenolic compounds is important, first to obtain better understanding of the relationships that could exist between these substances and the physiology and organoleptic qualities of the fruit, and secondly to provide a more solid basis for processing techniques, thus leading to improved quality. The major phenolic compounds in olive oil, identified and quantified by high performance liquid chromatography or capillary zone electrophoresis coupled with mass spectrometry, belong to four different classes: 1) simple phenols (hydroxytyrosol, tyrosol, vanillic acid, vanillin, p-coumaric acid, ferulic acid, 4-ethylphenol, tyrosyl and hydroxytyrosyl acetate) [9-12]; 2) polyphenols (flavonoids: apigenin and luteolin) [13]; 3) secoiridoids (oleuropein glucoside, aldehydic and dialdehydic forms of elenolic acid, of ligstroside and oleuropein aglycons, and the aglycone forms of elenolic acid, oleuropein [14-17], deacetoxyligstroside, ligstroside [11-12], dialdehydic form of decarboxymethyl elenolic acid linked to hydroxytyrosol and dialdehydic form of decarboxymethyl elenolic acid linked to tyrosol [18-19]), and 4) lignans ((+)-pinoresinol and (+)-1-acetoxypinoresinol) [20-22]). A recent detailed review describing current methological approaches for determination of individual phenolic compounds in virgin olive oil can be found in [23] and a brief overview of these compounds is given in table 1.

Table 1. Phenolic compounds in virgin olive oil

Category	Compound	References
Simple phenols	Hydroxytyrosol Tyrosol Vanillic acid Vanillin p-Coumaric acid Ferulic acid 4-Ethylphenol Tyrosyl acetate Hydroxytyrosyl acetate	[9-12]
Polyphenols	Flavonoids: apigenin and luteolin	[13]
Secoiridoids	Oleuropein glucoside Aldehydic form of elenolic acid Dialdehydic forms of elenolic acid Aldehydic form of ligstroside Dialdehydic forms of ligstroside Aldehydic form of oleuropein Dialdehydic forms of oleuropein Demethyloleuropein Oleuropein Deacetoxyoleuropein 10-Hydroxy-oleuropein Elenolic acid Ligstroside Deacetoxyligstroside Dialdehydic form of decarboxymethyl elenolic acid linked to hydroxytyrosol Dialdehydic form of decarboxymethyl elenolic acid linked to tyrosol	[11-12-14-19]
Lignans	(+)-Pinoresinol (+)-1-Acetoxypinoresinol	[20-22]

Virgin olive oil is produced by direct pressing or centrifugation of the olives. High-quality virgin olive oil is labelled and sold as extra-virgin olive oil. Virgin olive oils with high acidity must be chemically or physically refined, a process that results in lost of phenolic compounds. Commonly marketed “olive oil” is in fact a mixture of virgin and refined olive oil. After virgin olive oil production, the rest of the olive drupe and pit is further processed by chemical or physical extraction. After refining of the latter extract and once added to a certain quantity of virgin olive oil, the resulting mixture is sold as pomace olive oil [24]. Not surprisingly, the procedure of obtaining the olive oil is in itself a source of variability, the qualitative level of the percolated being superior to the centrifuged olive oil. Thus, virgin olive oils from percolation present: (i) higher content of total phenols, o-diphenols, hydroxytyrosol, tyrosol-aglycons, tocopherols, trans-2-hexenal, total volatiles and waxes; (ii) higher values of resistance to autoxidation and of turbidity; (iii) higher sensory scores; (iv) higher ratios of campesterol/stigmasterol, trans-2-hexenal/hexanal and trans-2-hexenal/total volatiles; (v) lower contents of chlorophylls, pheophytins, sterols and aliphatic and triterpene alcohols; (vi) lower alcoholic index and color indices; (vii) similar values of acidity, peroxide index and ultraviolet spectrophotometric indices; (viii) similar percentages of saturated and unsaturated fatty acids, triglycerides, and diglycerides; and (ix) similar values of glyceridic indices [25].

On the other hand, around 80% or more of the olive oil phenolic compounds are lost in the refining process [26]. Accordingly, the value of total phenols for extra-virgin (232 mg/kg, common range 130-350 mg/kg) was significantly higher than that of refined olive oil (62 mg/kg) as judged by high-performance liquid chromatography analysis. Tyrosol, hydroxytyrosol and their secoiridoid derivatives represent around 90% of the total phenolic content of a virgin olive oil. EVOO also contained higher concentrations of secoiridoids (27 mg/kg vs. 9 mg/kg) and lignans (41 mg/kg vs. 7 mg/kg) when compared to refined virgin olive oils [26].

The production of olive oils yields a considerable amount of waste-water, containing phenols and other natural antioxidants, that ends up as an environmental contaminant. The composition of olive oil waste-water depends on the extraction system, with the traditional press method resulting in decreased loss of antioxidant activity in this fraction. DellaGreca *et al.* [27] isolated and identified new phenolic compounds of olive oil mill waste-waters such as the lignan 1-hydroxy- 2- (4-hydroxy-3-methoxy phenyl)- 6- (3-acetyl-4-hydroxy-5-methoxyphenyl)- 3,7-dioxabicyclo [3.3.0] octane; the secoiridoid 2H-pyran- 4-acetic acid, 3-hydroxymethyl-2,3-dihydro-5-(methoxycarbonyl)-2-methyl-, methyl ester; the glycoside 4-[β-D-xylopyranosyl-(1-->6)]-β-D-glucopyranosyl-1,4-dihydroxy-2-methoxybenzene and the lactone 3-[1-(hydroxymethyl)-1-propenyl] delta-glutarolactone. As constituents of the waste-waters, known aromatic compounds were also present such as the acids: 4-hydroxybenzoic, protocatechuic, vanillic, 4-hydroxy-3,5-dimethoxybenzoic, 4-hydroxyphenylacetic, 3,4-dihydroxyphenylacetic, p-coumaric, caffeic, ferulic and sinapic; the alcohols: tyrosol, hydroxytyrosol, homovanillyl, 2-(3,4-dihydroxy)phenyl-1,2-ethandiol, catechol, 1-O-[2-(3,4-dihydroxy)phenylethyl]-

(3,4-dihydroxy)phenyl-1,2- ethandiol, 1-O-[2-(4-hydroxy) phenylethyl]-(3,4-dihydroxy) phenyl-1,2- ethandiol, D(+)-erythro-1-(4-hydroxy-3-methoxy)-phenyl-1,2,3-propantriol; and the glycosides: p-hydroxyphenethyl-β-D-glucopyranoside,2 (3,4-dihydroxyphenyl) ethanol 3β-D-glucopyranoside, and 2(3,4-dihydroxyphenyl) ethanol 4β-D-glucopyranoside. Some authors have stated that olive oil waste-water could be a good source of these compounds, which would contribute to diminish the environmental impact [28]. Indeed, olive waste-water has been employed in some studies, as will be discussed later in this chapter.

Table olives can also be considered a good source of phenolic intake, although the concentration depends on olive variety, maturation and the processing method. Regarding the first issue, Manzanilla variety had a higher concentration than Hojiblanca or Gordal [11]. In Chemlali olive, the total phenolic content varied from 6 to 16 g/kg expressed as pyrogallol equivalents. Its highest level was found at the last maturation period. Oleuropein, a secoiridoid glycoside esterified with a phenolic acid, was the abundant phenolic compound [29]. 5-O-caffeoilquinic acid, luteolin 7-O-glucoside and rutin were identified from Portuguese table olives [30]. In relationship with the maturation aspect, the concentration of hydroxytyrosol 4-beta-D-glucoside increased in olive pulp with maturation and could be the main phenolic compound in mature olives [31]. Lastly, the olive fermentation and the different storage methods are processes that could also affect the final olive phenolic composition. The individual evolution of phenolic compounds studied during the natural fermentation of black olives revealed that cyanidin 3-rutinoside and cyanidin 3-glucoside were the main anthocyanins identified in fresh olives, and they were not detected after one month of storage either in brine or in olive oil. The fruit colors were different when aerobic or anaerobic conditions were used, as a consequence of the different anthocyanin polymerizations that took place. At time zero, the biophenols observed in the olive juice were hydroxytyrosol-4-beta-glucoside, salidroside, verbascoside, oleuropein, hydroxytyrosol and tyrosol; after 12 months, the main phenol was hydroxytyrosol. The phenolic content in the oil phase of olives was also analyzed. The dialdehydic form of elenolic acid linked to hydroxytyrosol and tyrosol, oleuropein aglycon, and ligstroside aglycon were the main compounds found at the beginning of fermentation, but these were not detected after 3 months. The acid hydrolysis of the initial glucosides (in olive juice) and the aglycons (in oil phase) was, therefore, the main reaction that took place during fermentation [32]. In this way, olive paste storage increased 4-ethylphenol, hydroxytyrosol, hydroxytyrosol acetate, tyrosol and catechol, reaching their concentration up to 600 mg/kg of oil [10]. In other study, olives (*Olea europaea* cv. *Chondrolia*) were stored under a CO_2 atmosphere immediately after harvesting for a period of 12 days. Hydroxytyrosol glucoside, demethylated ligstroside aglycone, and verbascoside were found to be different in olives stored under CO_2 from those stored under regular atmospheric conditions [33]. Food processing is also a source of variability. Thus, olives not subjected to the Spanish brining process contain up to 16 g/kg of phenolic compounds typified by acteosides, hydroxytyrosol, tyrosol and phenyl propionic acids. Olives in brine had approximately 1.2 g/ kg whereas oxidized olives had the lowest (approximately 0.2 g/kg) [11]. Tyrosol, hydroxytyrosol, 3-(3, 4-dihydroxyphenyl) propanoic acid (dihydrocaffeic acid), dihydro-p-coumaric acid (phloretic acid), the phenylpropanoid glucosides acteoside (verbascoside) and isoacteoside, along with the flavonoids luteolin and apigenin are major components of the phenolic fraction of brined black olives. Brined green olives contained only hydroxytyrosol and traces of other minor phenols and showed even higher concentrations of phenolic antioxidants than olive oil [34]. Blekas *et al.* [35] analyzed the total phenol content in some commercial olives: Spanish-style green olives in brine, Greek-style naturally black olives in brine, and Kalamata olives in brine. Hydroxytyrosol, tyrosol and luteolin were the prevailing phenols in almost all of the samples examined. High levels of hydroxytyrosol were found mainly in Kalamata olives and Spanish-style green olives from cultivar Chalkidiki (250-760 mg/kg). Future research on olive and derivatives needs to be aware of these variables and provide unambiguous information regarding these aspects.

Chapter 3

AGRICULTURE AND PROCESSING MODIFY THE CONTENT AND TYPE OF OLIVE OIL PHENOLIC COMPOUNDS

The processes that influence the olive composition (e.g. variety, maturation, storage, plant irrigation strategy, extraction systems) will determine the quality of the olive oil obtained from those fruits. Several authors have analyzed the EVOO phenol content under these different conditions. There are many studies about the influence of the varietal plant on the phenolic content [16,29,36,38]. While total phenols content of Cornicabra extra-virgin olive oil was 75 mg/kg, the Picual content was 450 with intermediate values for Empeltre and Arbequina cultivars [39]. In addition, the type of phenol also changes. In fact, the five variables that most satisfactorily classified the principal commercial Spanish virgin olive oil varieties were 1-acetoxypinoresinol, 4-(acetoxyethyl)-1,2-dihydroxybenzene, ligstroside aglycon and the dialdehydic form of elenolic acid linked to tyrosol contents [18].

The concentration of phenols in EVOO decreased with ripeness of olive fruits [15-16-40]. The main change found was the increase of luteolin and the hydrolysis of the secoiridoid aglycons. This reaction gave rise to an increase in the free phenolics hydroxytyrosol and tyrosol in the oil [13]. In contrast, the concentration of lignans, 1-acetoxypinoresinol and pinoresinol, remained constant during storage. The total molar concentration of the phenolic compounds analyzed in the oils changed slightly (<20% reduction) after one year of storage [37]. The changes in phenols, pigments, and alpha-tocopherol with storage time followed first-order kinetics [41].

Plant irrigation strategy also changed the profile of the oil phenolic compounds [9] and greatly decreased the presence of secoiridoid phenolic derivatives of Cornicabra virgin olive oils [40]. Olive extraction systems used also had influence. Thus, oils produced by the continuous system contain comparatively higher levels of biophenols [36] and the effects of destoning of extra-virgin olive oils from olives of the Leccino, Moraiolo, Frantoio, Pendolino, Taggiasca, and Colombaia varieties on total secoiridoid content were variety-dependent [42].

Chapter 4

ORGANOLEPTIC PROPERTIES

Phenolic compounds also affect the organoleptic characteristics of the oil, and their degradation is one of the changes causing variation in taste through time. The bitter taste of virgin olive oil is essentially due to the dialdehydic and aldehydic forms of oleuropein and ligstroside aglycon [43]. In fact, Garcia *et al.* found a good correlation between oil bitterness and content of hydroxytyrosol secoiridoid derivatives [44]. Deacetoxy-ligstroside aglycon or oleocanthal produced a strong burning pungent sensation at the back of the throat [45]. In contrast, deacetoxy-oleuropein aglycon produced only a slight burning/numbing sensation, which was perceived more on the tongue [46]. A quantitative analysis in 10 selected throat-burning EVOO showed that the concentrations of oleacanthal, deacetoxy ligstroside aglycone, ranged from 22 to 190 μg/g [45]. Olive heating was found to promote a reduction of oil bitterness in direct relationship to the time and temperature used [44].

IN VITRO ANTIOXIDANT PROPERTIES AND SCAVENGING OF FREE RADICALS

Extra-virgin olive oil, obtained by physical pressure from a whole fruit, as previously mentioned, is rich in phenolic compounds that are responsible for the particular stability of the oil [47-48]. EVOO had the highest antioxidant potential and thermal stability compared with other vegetable oils, such as sunflower or corn oil. The antioxidant potential is the result of a combination of specific antioxidant phenols, tyrosol, hydroxytyrosol and oleuropein, and other phenolic compounds with radical scavenging abilities towards superoxide anion and hydroxyl radical. These compounds moreover act as effective stabilizers of alpha-tocopherol during olive oil heating [49], thus contributing to the nutritional value of cooked foods [50]. Other authors have found a correlation between antioxidant activity and total phenolic content of samples [29-30]. In fact, no individual compound was identified as the principal contributor to the overall antioxidant activity, and the total phenol determination was better correlated with antioxidant activity and oxidative stability, than individually-tested phenols or phenol groups such as o-diphenols [51]. Indeed, in a low content of phenolic compounds such as found in the Cornicabra virgin olive oil (38 mg/kg), the oxidative stability was predicted by taking into account either the sum of the dialdehydic form of elenolic acid linked to hydroxytyrosol and oleuropein aglycon (r = 0.91-0.96), or the sum of these two and hydroxytyrosol (r = 0.90-0.97) [18]. In other cases, a strong correlation between total phenols and antioxidant capacities was found, although not as high as that observed with beta-carotene. For this reason, the beta-carotene test was proposed as the best method of measuring the antioxidant capacity [52]. Globally, the stability of olive oils depends mainly on triacylglycerol composition and concentrations of o-diphenols and alpha-tocopherol [53]. At the same millimolar concentrations, the different o-diphenolic compounds yielded similar and significant increases of the induction time when they were subjected to oxidation, while tyrosol had a minor role [54]. Undoubtedly, this may represent a simplistic approach since flavonoids also showed antioxidant activities, having a hydroxyl group at the ortho position at 3' on the B ring of the flavonoid nucleus which could contribute to the antioxidant potential activity of the flavonoids [55]. In addition, other isolated phenolic compounds of EVOO (phenyl-ethyl alcohols, lignans and secoiridoids) also manifested antioxidant capacity [56]. Further support to this synergistic combination between antioxidants has been demonstrated by Franconi *et al.* using two virgin olive oils with variable minor polar phenolic content (Taggiasca oil of 208.5 mg/L and Seggianese oil of 441.9 mg/L), thus avoiding the instability of the phenol compounds out of their natural environment. The first oil was enriched in lignan derivatives free of 5-hydroxytyrosol and 5-hydroxytyrosol derivatives, whereas the major compounds in Seggianese oils were secoiridoid derivatives. They observed that the overall antioxidant potency correlated well with total phenolic content, Seggianese extract being more active [57]. This information is indicative of a synergistic combination between compounds with variable role depending on the specific composition of an olive oil.

Another degree of complication is provided by the natural instability of the compounds and their conservation, since the degree of oil degradation can alter the antioxidant activity. In fact, an EVOO with a low degradation level had a higher content of the dialdehydic form of elenolic acid linked to hydroxytyrosol and a lower content of hydroxytyrosol than oils having intermediate and advanced degradation levels. EVOOs with a low degradation degree were also 3-5 times more efficient as 2, 2-diphenyl-1-picrylhydrazyl radical scavengers and 2 times more potent as inhibitors of the superoxide-generating xanthine oxidase/xanthine reaction than oils having intermediate or advanced degradation levels [58].

In order to study the radical scavenging potencies of phenolic compounds, a wide range of tests have been employed. Several authors have used the 2, 2-diphenyl-1-picrylhydrazyl (DPPH) radical model to test the scavenging capacity of these phenolic substances. Thus, Perez-Bonilla [59], using purified hydroxytyrosol, tyrosol, cycloolivil, 7-deoxyloganic acid, oleuropein and ligustroside isolated from *Olea europaea* (Picual olive cultivar), found that hydroxytyrosol showed a higher activity than the natural antioxidant rosmarinic acid. Cycloolivil and oleuropein showed stronger activities than the synthetic antioxidant butylated hydroxytoluene against the same radical. Ligustroside, tyrosol and 7-deoxyloganic acid showed little activity. Hydroxytyrosol was a superior antioxidant and radical scavenger than oleuropein or tyrosol. All of these phenolic substances – including simple phenols (hydroxytyrosol, tyrosol), aldehydic secoiridoids, flavonoids and lignans (acetoxypinoresinol, pinoresinol) – were potent inhibitors of reactive oxygen species attack, on for example salicylic acid and 2-deoxyguanosine [60]. This author also quantified the activity of some hydroxytyrosol metabolites as antioxidants, using the DPPH test. While IC50 for homovanillic acid and homovanillic alcohol were 14.8 and 11.4 μM respectively, the 3-O-glucuronide conjugate was more potent than hydroxytyrosol, with a 2.3 in comparison to 11.0 μM, while the monosulfate conjugate was almost devoid of radical scavenging activity [61]. With more refinement in reaction mechanisms, it has been proposed that oleuropein, hydroxytyrosol and caffeic acid have the largest rate constants for the first H-atom abstraction values by the stable radical DPPH, whereas dihydrocaffeic acid, an intestinal metabolite of caffeic acid, was the best antioxidant. Regarding phenols, the presence of a catechol moiety has been proposed as responsible for the antioxidant effect [62]. An analysis of the main products formed in the induced oxidation of olive phenols revealed the presence of dimers and trimers. With hydroxytyrosol and dihydrocaffeic acid, oligomerization took place with the addition of water molecules. The olive phenols acted as retardants, rather than chain breakers like alpha-tocopherol. From a detailed mechanistic investigation, it appears that the inhibition of lipid peroxidation by olive phenols can be satisfactorily interpreted by assuming that they essentially reduce the 2,2'-azobis(2-amidinopropane) hydrochloride- derived initiating radicals. Overall, olive phenols prove to be efficient scavengers of hydrophilic peroxyl radicals with a long-lasting antioxidant effect owing to the residual activity of some of their oxidation products [63].

Using the model of oxidation initiated by 2,2'-azobis(2-amidinopropane) hydrochloride in a soybeanliposome system, the antioxidant activities of hydroxytyrosol, hydroxytyrosol acetate, oleuropein, 3,4-dihydroxyphenylelenolic acid and 3,4-dihydroxyphenylelenolic acid dialdehyde were similar. Synergistic effects were observed in the antioxidant activity of combinations of alpha-tocopherol with olive oil phenols both with and without ascorbic acid. Olive oil phenols did not penetrate into the membrane, but their effectiveness as antioxidants showed that they were associated with the surface of the phospholipid bilayer [64]. Synergistic properties of phenolic extracts rich in biophenols of EVOOs were additionally observed in a different model that tested the antioxidant activity of lactoferrin in lipid systems containing iron compared to alpha-tocopherol, ferulic or coumaric acids [65]. Antiradical and antioxidant activities of olive oil phenolic compounds were due mainly to the presence of a 3, 4-dihydroxy moiety linked to an aromatic ring, and the effect depended on the polarity of the phenolic compound. Glucosides and more complex phenolics exhibited higher antioxidant activities toward oxidation of liposomes, whereas in bulk lipids, aglycons were more potent antioxidants, with the exception of oleuropein. Lignans acted as antioxidants only in liposomes, which could partly be due to their chelating activity, since liposome oxidation was initiated by cupric acetate. The antioxidant activity of virgin olive oil was principally associated with the dialdehydic form of elenolic acid linked to hydroxytyrosol and luteolin [66]. The use of different approaches to determine antioxidant potency has been addressed by Stupans *et al.* [67], demonstrating that activity in one test does not necessarily correlate with activity in another. It was also shown that the presence of two phenolic groups is not always associated with antioxidant activity. We should thus be careful in the interpretation of results; for instance, the hydroxytyrosol was the most stable compound in the absence of Fe^{3+} but the most sensitive to its presence [68]. Overall, these *in vitro* approaches reinforce the data obtained in whole extra virgin olive of antioxidant activity but add a new level of complexity such as the exact mechanism studied in a particular test and the accessibility of compounds depending on the presence of nanomolecular barriers (e.g. phospholipid membrane).

COOKING METHOD MAY MODIFY THE INTAKE OF OLIVE BIOPHENOLS

Mounting evidence suggests that virgin olive oil has a high resistance to oxidative deterioration compared to other vegetable oils, as mentioned in the previous section. For this reason, olive oil has been used as preservative in the traditional Mediterranean cuisine, although only recent experiments have been addressed to study the influence of isolated phenolic compounds as well as the different procedures of adding the antioxidant (spraying, glazing or washing the fillets). Data on the efficacy of biophenols as food antioxidants has been compiled through numerous studies. Thus, biophenols were effective antioxidants when added to heated tuna muscle in the presence of either brine or refined olive oil, although the oxidation rate in tuna muscle packed in brine was higher than that of tuna packed in refined olive oil [70]. In the same sense, addition of 100 mg/kg of caffeic acid, hydroxytyrosol and propyl gallate could regenerate endogenous oxidation of alpha-tocopherol resulting in a reduced lipid oxidation in fish muscle supplemented with phenolic compounds [71]. To this purpose, the exogenous antioxidants have to be incorporated into the membranes where key substances for fish lipid oxidation are located [72].

The effects of olive oil can be altered by the cooking method. Since most cooking methods are based on heating the food with olive oil, the high temperatures (as well as oxygen and water absorption) cause triacylglycerols in the oil to undergo a series of reactions, namely hydrolysis, oxidation, isomerization and polymerization, and change antioxidant minor components along with the elimination of water. Experiments designed to mimic domestic cooking conditions undertaken by Brennes *et al.* showed that thermal oxidation of oils at 180 °C caused a significant decrease in hydroxytyrosol- and tyrosol-like substances. In contrast, oils heated for 25 h still retained a high proportion of the lignans 1-acetoxypinoresinol and pinoresinol. This thermal oxidation also resulted in a rapid degradation of alpha-tocopherol and the glyceridic fraction of oils. Nevertheless, microwaving of oils for 10 min caused only minor losses in phenols, and the oil degradation was lower than that in thermoxidation assays. Again, lignans were the least affected phenols and did not change during microwave heating. Boiling a mixture of virgin olive oil and water in a pressure cooker for 30 min induced the hydrolysis of the secoiridoid aglycons and the diffusion of the free phenols, hydroxytyrosol and tyrosol, from the oil to the water phase. Losses of biophenols were detected only at pH lower than 6. It is worth noting that all the heating methods assayed resulted in more severe phenolic losses and oil degradation for Arbequina than for Picual oil [73]. In conditions simulating common domestic deep-frying (10 min at 180 °C) using virgin olive oils, Gomez-Alonso *et al.* also observed a decreased concentration of hydroxytyrosol and its secoiridoid derivatives. After six frying operations, only about 10% of the initial components remained. The antioxidant activity paralleled this decrease. Furthermore, the concentration of polar compounds, oxidized triacylglycerol monomers (oxTGs), dimeric TGs and polymerized TGs rapidly increased from the sixth frying operation onward [74]. This deterioration during successive frying was similar to that observed by Andrikopoulos *et al.*, where phenolic species present in virgin olive oil ranged from 70-80% after the first frying, but dropped to 20-30% after the eighth frying. A further characterization of the most damaged species showed that tannic acid, oleuropein and the hydroxytyrosol-elenolic acid dialdeydic form had remarkable resistance in all frying sessions, while hydroxytyrosol and hydroxytyrosol-elenolic acid were more quickly eliminated. Pan-frying resulted in less phenolic species recoveries when compared with deep-frying after three to four frying sessions, which are most usual in the household kitchen [75]. Overall, these experiments are providing experimental evidence that the supply of phenols may be changed by the procedure of olive oil intake.

ANTIMICROBIAL AND ANTIVIRAL ACTIVITY

As traditional antibiotics become increasingly ineffective and as new, particularly viral, diseases remain intractable, the search for drugs derived from plants to be used in these arenas has been intensified in recent years. Since the use of plant products as antimicrobial agents has been extensively reviewed elsewhere [76], in this section we will focus on studies assessing antibiotic properties of the different compounds present in olive oil. In this aspect, hydroxytyrosol and oleuropein showed broad antimicrobial activity against several bacterial strains or clinical bacterial strains such as *Haemophilus influenzae*, *Moraxella catarrhalis*, *Salmonella typhi*, *Vibrio parahaemolyticus* and *Staphylococcus aureus* [77]. The potency was lower for oleuropein than for hydroxytyrosol. The latter also had antimycoplasmal activity against *Mycoplasma pneumoniae*, *Mycoplasma hominis*, and *Mycoplasma fermentans* [78]. When table olives were screened as potential antimicrobial agents, they also showed activity against Gram-positive (*Bacillus cereus, Bacillus subtilis, Staphylococcus aureus*) and Gram-negative (*Pseudomonas aeruginosa, Escherichia coli, Klebsiella pneumoniae*) bacteria and fungi [30]. Phenolic compounds also exerted a strong bactericidal activity against eight strains of *Helicobacter pylori*. Among these, the dialdehydic form of decarboxymethyl ligstroside aglycon showed the strongest bactericidal effect at a concentration as low as 1.3 ng/mL [79]. Like many antibiotics, olive oils showed bactericidal activity not only against harmful bacteria of the intestinal tract (*Clostridium perfringens* and *Escherichia coli*) and foodborne pathogens (*Listeria monocytogenes*, *Salmonella enterica*, *Yersinia* sp. and *Shigella sonnei*), but also against beneficial microorganisms such as *Lactobacillus acidophilus* and *Bifidobacterium bifidum*. The dialdehydic form of decarboxymethyl oleuropein and ligstroside aglycons, hydroxytyrosol and tyrosol, were the phenolic compounds that better correlated with bacterial mortality [80]. These results suggest olive oil as a promising antimicrobial agent for treatment of intestinal or respiratory tract infections, as well as dyspeptic disease.

Regarding antiviral activity, several experiments have been carried out. Thus, oleuropein and hydroxytyrosol were found to bind to the conserved hydrophobic pocket on the surface of the HIV-gp41 fusion domain by hydrogen bonds with Q577 and by hydrophobic interactions with I573, G572, and L568 on the gp41 N-terminal heptad repeat peptide N36, interfering in a dose-dependent manner with the HIV-1 fusion core formation, with EC(50)s of 66-58 nM, and exhibiting no detectable toxicity [81]. Oleuropein and hydroxytyrosol also exhibited dose-dependent inhibition on HIV-1 integrase activities including 3'-processing, strand transfer and disintegration, with EC(50)s in the nanomolar range [82]. These studies demonstrate a potential use of these compounds as anti-HIV agents.

BIOAVAILABILITY OF OLIVE OIL PHENOLICS

Our knowledge regarding the absorption and metabolism of phenols has advanced enormously in the past few years, since new and more efficient methods have been developed to monitor levels of the major phenolic compounds and their metabolites present either in plasma [83-84] or in urine [85]. Through these advances, it is becoming evident that although most biophenols are absorbed to some extent, this is very dependent on the type of biophenol.

The term 'bioavailability' describes that fraction of an administered dose of any olive oil metabolite that reaches the systemic circulation. Bioavailability of a dietary compound is hence a prerequisite for it to show any *in vivo* biological action. Furthermore, absorption may be accompanied by extensive conjugation and metabolism, and the forms appearing in the blood are usually different from those found in food. Based on these aspects, not all the effects of phenolic compounds observed *in vitro* may be automatically extrapolated *in vivo*. Thus, insight into the kinetics and metabolism of olive oil phenols is needed to assess its beneficial properties. To this end, in this section, we have reviewed the current knowledge on absorption, metabolism and urinary excretion of olive oil phenols *in vitro* and in animal and human studies.

Absorption. One of the main objectives of bioavailability studies is to determine, among the different dietary phenols, which of these compounds are better absorbed and which lead to the formation of active metabolites. As mentioned above, olives contain mainly the polar glycosides of oleuropein and ligstroside. Oleuropein and ligstroside are the parent compounds of the less polar oleuropein- and ligstroside-aglycones. Oleuropein- and ligstroside-aglycones are formed by removal of the glucose moiety from the oleuropein- and ligstroside-glycoside by β-glucosidase during ripening. A rapid oleuropein hydrolysis under gastric conditions has been described [86], resulting in significant increases in the amount of free hydroxytyrosol and tyrosol entering the small intestine. For glycosylated biophenols, the removal of the hydrophilic moiety will usually be necessary for passive diffusion across the small intestine brush border to occur. Therefore, the first step of metabolism should be removal of the sugar by glycosidases. Glycosidase activities can occur either in the olive itself or in the cells of the gastrointestinal mucosa.

Data obtained from *in vitro* experiments regarding the absorption of hydroxytyrosol and tyrosol seem to be in agreement. However, conflicting results have been described with regards to the absorption rates of oleuropein. Thus, the kinetic data provided by Manna *et al.*, using differentiated Caco-2 cell monolayers, demonstrated that hydroxytyrosol was quantitatively absorbed at the intestinal level via a passive bidirectional diffusion mechanism [87]. Results were confirmed by Corona *et al.*, who using *in vitro* and *ex vivo* techniques showed that both hydroxytyrosol and tyrosol were transferred across human Caco-2 cell monolayers and rat segments of jejunum and ileum. Oleuropein was rapidly degraded by colonic microflora resulting in the formation of hydroxytyrosol [86]. In contrast, Edgecombe *et al.* determined that oleuropein could be absorbed, albeit poorly, from isolated perfused rat intestine under both iso-osmotic and hypotonic luminal conditions using *in situ* intestinal perfusion technique. Under hypotonic conditions, the permeability of oleuropein was significantly greater due to an increase in the opening of paracellular junctions [88].

In animal studies, a hydrolyzed aqueous olive pulp extract with a limit dosage of 2 g/kg produced no toxicity in mice. In rats, an oral non-observed adverse effects level (NOAEL) of 2 g/kg was also established for a 90-day study [89-90]. Adverse effects were moreover absent in a rat developmental toxicity study. Isolated hydroxytyrosol did not show any toxicity up to 2 g/kg when orally administered to rats [60]. Using olive mill waste-water extract containing different amounts of hydroxytyrosol, Visioli *et al.* showed that hydroxytyrosol was rapidly and dose-dependently absorbed and this correlated with an increase in plasma antioxidant capacity [91]. Purified hydroxytyrosol and tyrosol were provided to rats to test the effect of different matrices. Thus, these compounds were administered intravenously (in saline) and orally (in oil- and water-based solutions). Oral bioavailability estimates of hydroxytyrosol, when administered in an olive oil solution and when dosed as an aqueous solution, were 99% and 75%, respectively. Bioavailability estimates of tyrosol, when orally administered in either an olive oil solution or an aqueous solution were 98% and 71%, respectively [92]. Similar figures were reported by Soni *et al.* for oral bioavailability of hydroxytyrosol in olive oil and in an aqueous solution [90]. These works provide experimental evidence on the influence of the matrix of phenolic compounds.

In human studies, a dose-dependent and non-saturable absorption of simple phenols has been shown, after either single dose or long-term exposure. Long-term studies have been tackled by Bonanome *et al.*, who fed fourteen healthy volunteers two diets, each for one month. The only difference between the diets was that the first supplied 50 g of EVOO/day, whereas the second one supplied 50 g of refined olive oil with no simple phenols. Unexpectedly, tyrosol and hydroxytyrosol were not recovered in lipoprotein. Conversely, Masella *et al.* reported that the plasma concentration of total phenolic compounds significantly increased from 11.3 mg/l after 6-week consumption of 20 g/day of olive oil with a low content of phenols (11 mg/kg), to 23.0 mg/l after 4-week consumption of 20 g/day of EVOO with a high content of phenols (238 mg/kg) [94]. In agreement with this, Miro-Casas *et al.* showed an increase in the absorption of tyrosol and hydroxytyrosol with moderate and sustained doses of virgin olive oil consumption, after both a single-dose ingestion (50 ml) and a short-term consumption (one week, 25 ml/day) of virgin olive oil [95]. Even in ileostomy subjects, olive oil phenols were absorbed [96]. Despite the discrepancy observed by Bonanome´s group regarding phenolic levels after chronic intake of olive oil, this group did observe that in healthy volunteers receiving an oral single fat load consisting of 100 g of EVOO, tyrosol and hdroxytyrosol were recovered in all lipoprotein fractions, except VLDL, with concentrations peaking between 60 and 120 minutes after feeding. Perhaps not unexpectedly, a high variability in tyrosol and hydroxytyrosol absorption was observed among subjects. Plasma antioxidant capacity increased significantly after 120 min, and then returned almost to baseline values after 360 minutes [93]. In this sense, Weinbrenner *et al.* showed a dose-dependent absorption of tyrosol and hydroxytyrosol when administered in oral fat loads of 25 ml olive oil with high, moderate and low phenolic content to twelve healthy male volunteers in a randomized, controlled, cross-over trial. Furthermore, the administered dose of 25 ml did not induce significant postprandial lipaemia, nor did it promote an increase of *in vivo* oxidation markers. With regard to plasma antioxidant enzymes, glutathione peroxidase activity decreased postprandially after low phenolic content olive oil ingestion; however this was not observed after intake of moderate and high phenolic content olive oils. The phenolic content of the olive oils administered may account for the protection of the endogenous antioxidant defences at postprandial state after ingestion of moderate and high phenolic content olive oils [97]. Covas *et al.* provided further confirmation in another study in which oral fat loads of 40 ml of similar olive oils, but with high (366 mg/kg), moderate (164 mg/kg) and low (2.7 mg/kg) phenolic content, were administered to 12 healthy male volunteers in a cross-over study design after a washout period. Tyrosol and hydroxytyrosol were found to be dose-dependently absorbed. Total phenolic compounds in LDL increased at postprandial state in a direct relationship with the phenolic compounds content of the olive oil ingested. Furthermore, the degree of LDL oxidation was lower as the phenolic content of the olive oil administered increased [98]. These findings suggest that phenolic compounds in olive oil are absorbed from the intestine, although not ileum, and may exert a significant antioxidant effect *in vivo*, probably in the postprandial phase.

Metabolism and excretion. Following absorption, phenolic compounds are conjugated by methylation, sulfation, glucuronidation or a combination of these. Such modifications alter the biological properties of the circulating metabolites. The biochemical pathways involved have been extensively reviewed elsewhere [99]. Like other drugs, the primary site of metabolism is determined by the dose: thus, large doses are metabolized primarily in the liver while small doses may be metabolized by the intestinal mucosa, with the liver playing a secondary role to further modify the phenolic conjugates from the small intestine [100]. *In vivo* and *in vitro* studies showed an extensive metabolization of hydroxytyrosol whilst tyrosol was transformed to a lesser extent.

In vitro studies undertaken by Mateos *et al.* detected methylated and glucuronidated forms of hydroxytyrosol at 18 h of hydroxytyrosol incubation with human hepatoma HepG2 cells. Hydroxytyrosyl acetate was largely converted into free hydroxytyrosol and subsequently metabolized, yet small amounts of glucuronidated hydroxytyrosyl acetate were detected. The same study showed poor metabolism of tyrosol, with <10% of the phenol glucuronidated after 18 h [101]. Using Caco-2 cells as a model of intestinal metabolism, hydroxytyrosol was converted in its O-methylated derivative, the product of catechol-Omethyltransferase [87]. Glucuronides of hydroxytyrosol and a novel glutathionylated conjugate of this phenolic compound were also detected [86]. Overall, these cell culture studies provide evidence that olive oil phenols are metabolized by the liver and intestine.

Since reliable data on plasma concentration of olive oil phenols are scarce, an alternative has been to look at olive oil phenols excreted in urine. The amount and form in which the olive oil phenols are present in urine may also give insight into their metabolism in the body. Different pharmacokinetic studies assessing phenol metabolites have been performed in animal models. Del Boccio *et al.* measured plasma concentrations of oleuropein and hydroxytyrosol in rats fed an oral single dose of oleuropein (100 mg/kg). They found the presence of unmodified oleuropein in plasma, which reached a peak value of 200 ng/ml within 2 h, with a small amount of hydroxytyrosol. In urine, both compounds were mainly found as glucuronides [102]. Tuck *et al.* administered [^{14}-C] hydroxytyrosol to rats, and 90% of the radioactivity was excreted in urine collected up to 5 h after injection, with about 5% detectable in feces and gastrointestinal content. Furthermore they found that hydroxytyosol was renally excreted either unchanged or as its metabolites such as glucuronide conjugate, sulfate conjugate, homovanillic acid, homovanillic alcohol, 3,4-dihydroxyphenylacetic acid and 3,4-dihydroxyphenylacetaldehyde. They also observed that tyrosol was excreted unchanged and as its conjugates [60]. For both compounds, the intravenously and orally administered oil-based doses resulted in significantly greater elimination of the phenolics in urine within 24 h than the oral, aqueous dosing method. There was no significant difference in the amount of phenolic compounds eliminated in urine between the intravenous dosing method and the oral oil-based dosing method for either tyrosol or hydroxytyrosol. [92]. Although these studies do shed some light onto the kinetics field, they also raised some criticism when Visioli *et al.* stated that the rat is not an appropriate model for the study of hydroxytyrosol metabolism since its urinary excretion was greater in humans [103].

Nevertheless, when compared to animal data, human studies show a similar presence of plasma metabolites. For instance, Torre-Carbot *et al.* were able to detect hydroxytyrosol monoglucuronide, hydroxytyrosol monosulfate, tyrosol glucuronide, tyrosol sulfate and homovanillic acid sulfate in low-density lipoprotein fractions and their glucuronide conjugates in urine [104]. A similar excretion profile was detected in 24-hour urine collections of healthy volunteers after administration of virgin olive oil by Caruso *et al.*, who showed that hydroxytyrosol underwent the action of catechol-O-methyl transferase resulting in an enhanced excretion of homovanillic alcohol and a significant increase of homovanillic acid, indicating an oxidation of the ethanolic residue of hydroxytyrosol and/or of homovanillic alcohol in humans through alcohol dehydrogenase and aldehyde dehydrogenase [105]. The authors also pointed out that excretion of both metabolites significantly correlated with the dose of administered hydroxytyrosol [106], in agreement with previous observation made by Visioli *et al.* in which hydroxytyrosol was found excreted in the urine as glucuronide conjugates and an increase in the administered dose increased the proportion of conjugation with glucuronide [107].

Miro-Casas *et al.* have extensively studied olive oil metabolites in different human trials. They showed an increase in 24 h-urine of tyrosol and hydroxytyrosol in seven healthy volunteers after both a single-dose ingestion (50 ml) and short-term consumption (one week, 25 ml/day) of virgin olive oil. Urinary recoveries for tyrosol were similar after a single dose and after sustained doses of virgin olive oil. However, values for hydroxytyrosol after sustained doses were 1.5-fold those obtained after a single 50 ml dose [95]. This group has also investigated hydroxytyrosol and tyrosol urinary recovery kinetics in response to olive oil administration, obtaining maximal peaks of excretion in the first 4 h. A further characterization of these metabolites indicated that hydroxytyrosol and tyrosol were mainly excreted in conjugated form, since only 5.9 % (hydroxytyrosol) and 13.8 % (tyrosol) of the total amounts excreted in urine were in free form [108]. When this group went through another kinetics study, they found increased plasma hydroxytyrosol and 3-O-methyl-hydroxytyrosol levels as a response to 25 ml of virgin olive oil administration, reaching maximum concentrations at 32 and 53 min, respectively. The estimated hydroxytyrosol elimination half-life was 2.43 h. Free forms of these phenolic compounds were not detected in plasma samples. Approximately 98% of hydroxytyrosol appears to be present in plasma and urine in conjugated forms, mainly glucuronoconjugates, suggesting extensive first-pass intestinal/hepatic metabolism of the ingested hydroxytyrosol [109]. Vissers estimated that the plasma concentrations after consuming 50 g of olive oil with 2 mg of hydroxytyrosol per day would reach 0.06 μmol/ l [110].

To sum up the above findings, it is worth noticing that although olive oil has been proved as beneficial regarding several biomarkers of cardiovascular risk in a large number of recent intervention studies, olive oil is made up of a mixture of different biophenols and the exact nature of the most active compounds is still unclear. Absorption, metabolism and elimination vary widely between biophenols. These data on bioavailability should be thus taken into account to improve the experimental design and the interpretation of the observed effects. Future intervention studies should include a detailed assessment of the bioavailability of phenolic compounds [111].

Chapter 5

PHENOLIC COMPOUNDS OF VIRGIN OLIVE OIL ARE EFFECTIVE ANTIOXIDANTS IN *EX VIVO* AND *IN VIVO* STUDIES

Olive oil phenolic compounds have been clearly proved to be antioxidants and free radical scavengers *in vitro*, but their effect in cellular and animal models should be also demonstrated. To this purpose, the ability to scavenge or depress radical oxygen species (ROS) has been evaluated using cellular models. Thus, hydroxytyrosol inhibited luminol-amplified chemiluminescence of human neutrophils stimulated with N-formyl-methionyl-leucyl-phenylalanine, phorbol myristate acetate or opsonized zymosan. The effect was dose-dependent and occurred immediately after the addition of the phenol. However, hydroxytyrosol had no effect on lucigenin-amplified chemiluminescence, suggesting that it did not inhibit NADPH oxidase activation or scavenge superoxide anions. Furthermore, hydroxytyrosol inhibited H_2O_2-dependent-dichlorofuoroscein fluorescence of activated neutrophils. Finally, hydroxytyrosol inhibited luminol-amplified chemiluminescence in a cell-free system consisting of H_2O_2 and peroxidase. These results indicated that hydroxytyrosol could exert its antioxidant effect by scavenging hydrogen peroxide but not superoxide anion (O_2^-) released during the respiratory burst [112]. The latter effect may be different according to the experimental approach since other authors have found effective scavenging activity against superoxide anion (O_2^-) [113] and hydroxyl radical ($OH^·$), an effect attributed to the catechol moiety [114]. Hydroxytyrosol may not be the only active molecule present in olives regarding the scavenging activity. Thus, Leger *et al.* [115] using olive oil waste water fractions added to the human monocytic line THP-1, observed lowered levels of O_2^- due to the ability to scavenge and depress superoxide production, and oleuropein was found as a potent scavenger of superoxide radicals and inhibitor of neutrophils respiratory burst [113].

Due to the growing evidence that ROS are involved in the aetiology of several colorectal pathologies, the effects of phenolic compounds have been also addressed on intestinal cell lines such as Caco-2. In this regard, Manna *et al.* [116] showed that hydroxytyrosol completely prevented the H_2O_2-induced alterations in these cells. In contrast, tyrosol was ineffective. Initially, these effects were attributed to the antioxidant activity due to the ortho-dihydroxy moiety of the molecule. Posterior assays of chemically stable acetyl derivatives of the natural hydroxytyrosol, modified in the o-diphenolic ring and devoid of any chemical antioxidant activity, resulted at micromolar concentrations in equal protection against tert-butylhydroperoxide- induced oxidative damages in Caco-2 cells [117], suggesting a more elaborated mechanism may be involved.

The protection against lipid peroxidation has been a matter of study as well. In this aspect, hydroxytyrosol showed a substantial degree of inhibition against ascorbate-Fe^{2+} induced lipid peroxidation in rat liver microsomes that was at least 6 times higher than that observed in the presence of oleuropein. Inhibition of lipid peroxidation by tyrosol was not observed. In rat liver microsomes incubated alone or in the presence of tyrosol, the fatty acid composition was profoundly modified when subjected to *in vitro* peroxidation mediated by ascorbate-Fe^{2+}, with a considerable decrease of 18:2n6 and 20:4n6. In comparison, changes in fatty acid composition were not observed when microsomes were incubated with hydroxytyrosol. When oleuropein was used at low concentration (5, 15 μM) a considerable decrease of 20:4n6 was observed, but 18:2n6 was not modified; at higher concentration (30, 60 μM) changes in fatty acid composition were not observed. There was a very good correlation between the presence of oxidized phospholipids and the changes in polyunsaturated fatty acids previously observed. Thus, hydroxytyrosol showed the highest protection again oxidized phospholipid formation. The presence of oleuropein at low concentration (5, 15 μM) did not prevent the formation of oxidized phospholipids, although concentrations higher than 30 μM avoided completely the formation of these molecules. Tyrosol at any concentration assayed was found to be ineffective and allows the formation not only of oxidized phospholipids but also of oxidized cholesterol [118].

The protection against nitrogen reactive species has been equally considered. Thus, caffeic acid, oleuropein and hydroxytyrosol were active as scavengers of nitric oxide generated spontaneously from the decomposition of sodium nitroprusside (approximately 50% inhibition achieved at 75 μM), and had similar ability to scavenge chemically-generated peroxynitrite. Tyrosol was less active in these tests, probably because it lacks the catechol moiety. Despite their ability to interact with chemically prepared nitric oxide, neither oleuropein nor hydroxytyrosol at 5 μM altered nitric oxide- mediated relaxations of a nerve-stimulated rat anococcygeus preparation, suggesting that the nitrergic transmitter is protected from the effects of externally applied scavengers [119]. Hydroxytyrosol was also found highly protective against the peroxynitrite-dependent nitration of tyrosine and DNA damage by peroxynitrite *in vitro* [120]. Collectively, these data suggest that phenolics possess ability to scavenge both reactive oxygen and nitrogen species, but their impact may be restricted to those species present in the extracellular environment [119]. Concordant with this interpretation, the antioxidant phenolic compounds present in olive oil were potent inhibitors of the abundant free radical generation generated by the fecal matrix [121].

More elaborate models have been incorporated to demonstrate antioxidant properties of olive oil constituents. For example, the protective effect of oleuropein against post-ischemic oxidative burst was investigated by measuring the release, in the coronary effluent, of oxidized glutathione as a sensitive marker of oxidative stress. Pre-treatment with oleuropein before ischemia resulted in a significant decrease in creatine kinase and reduced glutathione release in the perfusate. Re-flow in ischemic hearts was accompanied by a prompt release of oxidized glutathione; in ischemic hearts pre-treated with oleuropein this release was significantly reduced. Membrane lipid peroxidation was also prevented by oleuropein suggesting a direct cardioprotective effect of oleuropein in the acute events that follow coronary occlusion [122]. Surprisingly, in studies using rabbit heart mitochondrial membranes, the saponifiable fraction of the experimental diets proved a more important role in preventing lipid peroxidation than the unsaponifiable fraction [123].

Visioli *et al.* [124] studied the effect of olive oils in humans, with increasing concentrations of hydroxytyrosol and oleuropein aglycone. They report a dose-dependent reduction in the urinary excretion of 8-iso-$PGF_{2\alpha}$, a biomarker of oxidative stress-induced *in vivo* lipid peroxidation. The statistically significant negative correlation between 8-iso-$PGF_{2\alpha}$ and homovanillyl alcohol urinary concentrations suggested that this hydroxytyrosol metabolite better reflected the *in vivo* activities of hydroxytyrosol. Likewise, the beneficial effect of olive oil phenols, especially hydroxytyrosol, in preventing lipid peroxidation was also demonstrated in rats exposed to passive smoking. This oxidative insult increased the urinary excretion of 8-iso-$PGF_{2\alpha}$. Treatment with the olive mill waste-water extract (5 mg/kg, providing 414 μg/kg of hydroxytyrosol) was able to completely prevent the increase at 48 hours and resulted in lower increments of 8-iso-$PGF_{2\alpha}$ excretion at 96 hours. Regarding the oxidative process reflected by this marker, the administration of phenolic compounds may be useful to reduce its biological impact.

A recent study analysed protection against oxidative DNA damage in healthy human subjects consuming olive oils with different concentrations of natural phenols. The oxidative DNA damage during treatment with high-phenol content oil was 30 % lower than the average during the low-content treatment. Urinary excretion of hydroxytyrosol and its metabolite homovanillyl alcohol were significantly increased in subjects consuming high-phenol content oil. This interesting finding of a reduction in DNA damage by consumption of an EVOO rich in phenols needs further confirmation due to the low number of subjects studied [125].

To avoid the instability of the phenolic compounds out of their natural environment and to improve their biological properties, new synthetic compounds are being developed. In this regard, the olive oil phenol hydroxytyrosol, as well as its metabolite homovanillic alcohol, were subjected to chemoselective lipase-catalyzed acylations. Hydroxytyrosol, homovanillic alcohol and their lipophilic derivatives showed very good DPPH radical scavenging activity. When these compounds were subjected to an atypical Comet test on whole blood cells, hydroxytyrosol, and its analogues with little hydrophobic character, showed a good protective effect against H_2O_2 induced oxidative DNA damage [126]. In another approach, hydroxytyrosol showed a protective effect against oxidation of proteins and lipids caused by peroxyl radicals, using a brain homogenate as an *ex vivo* model. Inhibition of the generation of carbonyl groups in protein was greater than of malondialdehyde in lipids [127]. These results prove that the inclusion of a lipophilic chain in the hydroxytyrosol molecule enhances its protection against oxidative damage and constitutes a promising beginning to further developments.

Chapter 6

Prevention of Low-Density Lipoprotein Oxidation by Olive Oil Phenolics

Oxidized LDL (OxLDL) is a proatherogenic lipoprotein, accumulating in the vascular wall and contributing to the pathogenesis of vascular dysfunction early in the development of atherosclerosis. Enhanced serum levels of OxLDL, as well as antibodies against its epitopes, are predictive for endothelial dysfunction and coronary heart disease. While enhanced oxidative stress is one factor triggering formation of modified LDL, OxLDL itself has been identified as a potent stimulus for vascular oxygen radical formation. This triggers a vicious circle with consequent endothelial dysfunction, tissue remodelling and vasospasm [128]. In this scenario, olive oil phenolic compounds acting as antioxidants that would prevent LDL oxidation have been proposed as potential agents in cardiovascular prevention. Several research works have been carried out in this direction. To start with, Leenen *et al.* [129] proposed an experimental design that attempts to mimic the *in vivo* situation. They incubated plasma with either individual olive oil phenols, or olive oil extracts with different phenolic compositions, and LDL was subsequently isolated and challenged for its resistance to oxidation. Results showed that the ortho-dihydroxy phenols (hydroxytyrosol and oleuropein-aglycone) were more efficient than their mono-hydroxy counterparts (tyrosol and ligstroside-aglycone) in increasing the resistance of LDL to oxidation [130]. However, the concentration of antioxidants required to inhibit LDL oxidation when added to whole plasma was substantially higher as compared to previous data where antioxidants were added directly to isolated LDL. In conclusion, this study supports the hypothesis that EVOO phenols in plasma protect LDL against oxidation [129]. Other *in vitro* experiments also demonstrated that hydroxytyrosol inhibited the production of isoprostanes and other markers of lipid peroxidation during LDL oxidation [131]. More recently Briante *et al.* have described that copper ion-induced LDL oxidation was inhibited by oleuropein and hydroxytyrosol in the initiation phase of the reaction at concentrations of phenols higher than that of Cu^{2+} ions; but, unexpectedly, at lower concentrations, both phenols anticipated the initiation process of LDL oxidation, thus exerting pro-oxidant capacities. The net effect of oleuropein and hydroxytyrosol on Cu^{2+}-induced LDL peroxidation would thus be determined by a balance of their pro- and antioxidant capacities. It is relevant to emphasize that this paper also described a synergistic effect among phenolic compounds that enhanced their antioxidant capacities, thereby avoiding the initial pro-oxidant effects during Cu^{2+}-induced LDL oxidation [132]. Homovanillic alcohol was a significantly more potent antioxidant than the other phenolics such as oleuropein, tyrosol, and hydroxytyrosol, both in chemical assays and in prolonging the lag phase of LDL oxidation [133]. In agreement with these findings, the minor phenolic and non-phenolic constituents of olive oil were found, in optimal doses of 10 μM or 20 μM, to have remarkable biological activity against copper ion-induced LDL oxidation, contributing to that previously reported for the major phenolic compounds. The main phytosterols, beta-sitosterol, campesterol, and stigmasterol, were found to have 43.8, 37.3, and 33.4% LDL mean protection activity, respectively. The triterpenoid derivative compounds, ursolic acid, uvaol and oleanolic acid, had similar protection activities of 50.5, 46.8, and 46.0%, respectively. The flavonoids, quercetin, luteolin, and rutin, exhibited the highest protection activities: 46.8, 49.5 and 53.7%, respectively, comparable to the 49.0% activity found for oleuropein. Overall, these findings indicate the relative independence of LDL protection activity with regard to structural differences among the compounds involved [134]. Further support to the contribution of minor phenols on the antioxidant activity toward LDL was provided by Romani *et al.* [135] working with Olivastra Seggianese EVOO where, minor polar compound content, particularly deacetoxy-oleuropein aglycone, was correlated with the antioxidant activity on human LDL. In experimental rabbits fed diets with similar vitamin E but different oil biophenol content (e.g. oleic acid-refined olive oil, EVOO and high oleic sunflower seed oil), Wiseman *et al.* [136] observed that antioxidants present only in EVOO (hydroxytyrosol and tyrosol) may contribute to the endogenous antioxidant capacity of LDL. The protective effect of olive oil phenols on models of minimally oxidized LDL has likewise been investigated. Thus, tyrosol and hydroxytyrosol inhibited J774 A1cell-mediated oxidation of LDL but to a different extent (100% hydroxytyrosol vs. 40% for tyrosol), and counteracted the impairment of antioxidant cellular defence, i.e. GSH and related enzymes. Tyrosol was effective in inhibiting about 30% of

ROS production only at later time-points (12 h for superoxide, 24 h for hydrogen peroxides). Interestingly, both phenols were effective when added to the cells for 2 h and removed before LDL treatment. This was probably related to cell-phenol interactions: hydroxytyrosol was rapidly found inside the cells and disappeared within 18h, while tyrosol accumulated intracellularly with time. This work hypothesized that in spite of its weak antioxidant activity, tyrosol was effective in preserving cellular antioxidant defences, probably by intracellular accumulation [137]. Collectively, all these results indicate that olive oil antioxidants increase the resistance of low-density lipoproteins against oxidation.

Current studies have been focused to gain a better understanding of the potential health benefits that EVOO could provide. In one of these studies the analysis was conducted on 28 outpatients older than 50 years. The supplement was taken for 6 weeks and involved daily consumption of 2 tablespoons (approx. 20 g) of EVOO as part of the subjects' habitual diets, without changing their usual total energy and fat consumption. This phase was followed by a 2-month wash-out. After dietary supplementation with olive oil, serum total- and LDL-cholesterol mean concentrations were lowered by 0.818 and 0.782 mmol/l, respectively. There was also a significant decline in the total-to-HDL and LDL-to-HDL cholesterol ratios. Linoleic and arachidonic acid content in serum phospholipids decreased significantly. These results point to an overwhelmingly beneficial influence of olive oil on the lipoprotein spectrum. The statistically significant reduction in arachidonic acid may even be regarded as clinically desirable [138]. Likewise, in a cross-over study of three randomly assigned sequences of daily administration of 25 ml of 3 different olive oils with low (2.7 mg/kg), medium (164 mg/kg) or high (366 mg/kg) phenolic content for a 3-week intervention period, preceded by 2-week washout periods, a linear increase in high-density lipoprotein (HDL) cholesterol levels was observed for low-, medium-, and high-phenol olive oil; mean change: 0.025, 0.032 and 0.045 mmol/l, respectively. Total cholesterol-HDL cholesterol ratio decreased linearly with the phenolic content of the olive oil. Triglyceride levels decreased by an average of 0.05 mmol/l for all olive oils. Oxidative stress markers decreased linearly with increasing phenolic content. Mean changes for oxidized low-density lipoprotein levels were 1.21, -1.48 and -3.21 U/l for the low-, medium- and high-phenol olive oil, respectively. The phenolic content of olive oil provide benefits for plasma lipid levels and oxidative damage in human subjects [139], corroborating previous pilot interventions of the group [140]. In an independent study, the lag time of LDL oxidation from the subjects consuming the Mediterranean diet was greater than those with the carbohydrate diet [141]. But in spite of precedent results, some human intervention studies have failed to demonstrate similar consistent effects. The explanation might be that the dose of the phenols, and thus their plasma concentration, was too low to influence *ex vivo* OxLDL levels [129]. A similar concern has been raised by Vissers *et al.* [110] although in stronger terms, since they even stated that the available evidence does not suggest that consumption of phenols in the amounts provided by dietary olive oil will protect LDL against oxidative modification to any important extent. More recently, Lapointe *et al.* [142] have reviewed inconsistent documented effects of flavonoid-rich food in humans, explaining them by difference in variety and quantity of flavonoid compounds used among studies. However, these authors proposed that a healthy food pattern such as the Mediterranean diet, which includes a combination of antioxidant compounds and flavonoid-rich foods, appears effective in decreasing OxLDL, which could give some mechanistic insight into the cardiovascular benefits associated with the Mediterranean diet. In conclusion, the hypothesis that olive oil phenolic compounds protect LDL in plasma against oxidation is perhaps still open, albeit a new dimension should be added to understand the concerted chemical cooperation of all phenolic and non-phenolic compounds present in extra virgin olive oil.

Chapter 7

Phenolic Compounds Offer an Indirect Protection by Activating Endogenous Cellular Defence Systems

Olive oil phenols have demonstrated their therapeutic potential by their direct antioxidant capacity, acting as radical scavengers and protecting LDL against peroxidation. Nevertheless, most recent experimental data indicate that phenols may offer an indirect, but possibly more robust, protection by activating endogenous defence systems. As proposed by Kaliora *et al.* [143], the knowledge of genetic factors involved in oxidation may open new horizons to identify persons at risk for cardiovascular disease, allowing effective dietary intervention strategies to recover normal homeostasis and to prevent diet-related implications. On this basis, current studies on the action of selected antioxidant compounds (olive oil phenolic compounds) on the activity of transcription factors, such as final targets in the signal transduction cascade and gene regulation, may emerge as new treatment concepts.

Hydroxytyrosol and other phenolic compounds, present in olives or virgin olive oil, have shown a protective effect against lipid peroxidation at cellular level in different models of free radical production, such as: 1) cadmium intoxication [144], 2) H_2O_2- or tert-butylhydroperoxide- induced oxidative damage [145-147], 3) lipid peroxidation observed in rats fed cholesterol- enriched diets [148], and 4) cyclosporine-induced ROS formation [149]. These protective effects are due in part to their activity as ROS and radical nitrogen species (RNS) scavengers but also by activating intracellular defence systems. In this way, in the first experimental approach, Casalino *et al.* [144] described that hydroxytyrosol induced glutathione synthesis in cadmium-intoxicated rats, contributing to protection against cellular lipid peroxidation. In the second approach, Manna *et al.* [145] also attributed to this molecule the capacity of preventing lipid peroxidation and oxidative hemolysis in erythrocytes secondary to H_2O_2-induced stress. In human prostate cancer cells injured with hydrogen peroxide, hydroxytyrosol at low concentrations (10 μM) led to lower levels of hydroperoxides, DNA damage and mRNA levels of glutathione peroxidase. Tyrosol reduced DNA oxidation only at high (>50 μM) concentrations and increased glutathione peroxidase (GPx) and GPx mRNA levels. Therefore, hydroxytyrosol was proposed to be the only significant active phenolic in comparison with tyrosol, the number and position of hydroxyl groups being once again crucial for the cellular effects of hydroxytyrosol [146]. The cytoprotective effect of hydroxytyrosol against tert-butylhydroperoxide treatment has been corroborated by other authors in cultured hepatic cells, in association with an increase in glutathione levels [147]. Likewise, in the third group of experiments proposing a nutritional model of oxidative stress, Fki *et al.* [148] have demonstrated that both hydroxytyrosol and olive mill waste-waters (OMW) extract were able to lower the thiobarbituric acid-reactive substances in liver, heart, kidney and aorta as compared with those of rats fed a cholesterol-rich diet. In addition, OMW phenolics increased catalase and superoxide dismutase activities in liver, enhancing total antioxidant enzyme activity. Finally, using a pharmacological ROS induction model, Galletti *et al.* [149] described that incubation of immortalized rat tubular cells with cyclosporin (CsA) induced a significant decrease in cell viability paralleled by intracellular ROS formation and alterations in lipid peroxidation. There was also an imbalance of glutathione redox state as well as up-regulation of heme oxygenase-1 (HO-1). Hydroxytyrosol, at micromolar concentration, quantitatively prevented the ROS formation and membrane lipid peroxidation. Furthermore, hydroxytyrosol strengthened the CsA induction of HO-1 expression. Considering that up-regulation of HO-1 has been associated with marked cytoprotection [150], this supposes a new step in the protecting role of this olive phenol against cellular aggression.

More cellular and tissue defence mechanisms are being explored [151], [152]. Hydroxytyrosol and other phenolic compounds present in virgin olive oil, such as oleuropein glycoside and gallic acid, have demonstrated their capacity to inhibit androstenedione 6beta-hydroxylase (CYP3A marker in human liver microsomes), postulating the 3, 4-dihydroxyphenyl ethanol structure was required for the inhibition. Considering that CYP3A has been associated with bioactivation of procarcinogens [153], its observed reduction induced by hydroxytyrosol could represent a potential target in cancer prevention and another aspect of cellular defence. Other possible action of olive phenols could be via increasing paraoxonase1 (PON1) activity. In fact, polyphenols increased PON1 mRNA expression and activity by an aryl hydrocarbon receptor-dependent mechanism. Beyond its antioxidative properties, PON1 prevented macrophage foam cell formation, attenuation of cholesterol and oxidized lipid influx, inhibition of macrophage cholesterol biosynthesis and stimulation of macrophage cholesterol efflux. These effects collectively confer additional anti-atherogenic properties [154].

These studies represent our endeavour to obtain more evidence on novel specific biological effects of olive phenolics that are as yet poorly understood, as well as the molecular mechanisms involved. As yet another example, hydroxytyrosol and tyrosol have been found to increase intracellular calcium in a dose-dependent manner [155]. Recent research on plant-derived polyphenols as potential anti-ageing nutraceutical agents has placed its attention on the transcription factors belonging to the forkhead/winged helix box gene, group O (FOXO) subfamily. FOXO transcription factors have been found to be crucial in downstream suppression of the life-shortening effects of insulin/insulin-like growth factor-I receptor signalling pathways that, when upregulated, accelerate ageing by suppression of FOXO. The various adverse processes activated upon FOXO suppression include increased generation of reactive oxygen species. ROS are pivotal for the onset of various common conditions, including hypertension, atherosclerosis, type-II diabetes, cancer and Alzheimer's disease, each of which shortens lifespan. In humans, FOXO3a, as well as FOXO1 and -4, and their downstream effectors, could hold the key to counteracting ageing and common diseases. An understanding of the processes controlled by these FOXOs should permit development of novel classes of agents that will more directly counteract or prevent the damage associated with diverse life-threatening conditions, and so foster a life of good health to a ripe old age. Just like caloric restriction, lifespan can be increased in various species by plant-derived polyphenols, such as resveratrol, via activation of cell sirtuins [156]. Sirtuins, such as SIRT1 in mammals, utilize FOXO and other pathways to achieve their beneficial effects on health and lifespan [156]. Moreover, activation of SIRT1 can protect cells also by the promotion of DNA repair capacity [157]. Braam *et al.* [158], using cultured human fibroblasts, monitored the transcriptome in order to evaluate how aging-associated oxidative stress influences gene expression. Their results showed that gene expression of fibroblasts was affected by donor age and a subset was corrected by antioxidants. Hydroxytyrosol treatment modulated a similar number of genes and induced cell-cycle regulatory genes. Overall, these findings indicate that gene expression in response to oxidative stress needs several cascades to activate the process and that many genes may be driven in this action.

Several lines of evidence also suggest a close connection between exogenous and endogenous antioxidants that appear to act in a synergistic fashion [159]. In an experimental design, already mentioned in this chapter, as is the minimally-oxidized LDL obtained by incubation with murine J774 A.1 macrophage-like cells, both protocatechuic acid and oleuropein interact with the following effects: 1) completely prevented the J774 A.1-mediated oxidation of LDL; 2) counteracted the time-dependent variations in intracellular redox balance, inhibiting the production of superoxide and H_2O_2 and the decrease in glutathione (GSH) content; 3) restored glutathione reductase (GR) and GPx activities; and 4) restored the mRNA expression of glutamylcisteine synthetase (γ-GCS), GR and GPx to control values. Their findings suggested that the activation of mRNA transcription of GSH-related enzymes represents another powerful mechanism in phenolic antioxidative action [160]. The latest studies strongly indicate that dietary biophenols can stimulate antioxidant and detoxification defence systems through antioxidant responsive elements present in the promoter regions of many of the genes inducible by oxidative and chemical stress [159].

Another aspect, already raised in this chapter, is that the effect of olive oil may not be simplisticly explained by a single compound. This has been tackled by Faine *et al.* [161] who studied the effect of EVOO, oleic acid or hydroxytyrosol administered separately to male Wistar rats. This work demonstrates that extra-virgin olive oil was more effective than its principal isolated components (oleic acid or hydroxytyrosol) in improving lipid profile, elevating high-density lipoprotein and diminishing low-density lipoprotein cholesterol concentrations. Olive oil also induced decreased antioxidant manganese superoxide dismutase activity and diminished protein carbonyl concentration, indicating that olive oil may exert direct antioxidant effect on myocardium. Hydroxytyrosol induced elevated aerobic metabolism, triacylglycerol and lipid hydroperoxide concentrations in cardiac muscle, indicating that long-term intake of this phenol may induce undesirable pro-oxidant activity on myocardium. In our opinion, the confirmed pro-oxidant activity on myocardium of hydroxytyrosol should advise against its use alone. On the other hand, this work confirms the cooperation among all the phenolic and other biologically-active compounds present in olive virgin oil to exert the benefits of virgin olive oil regarding endogenous antioxidant defence.

Chapter 8

ANTI-PROLIFERATIVE AND PRO-APOPTOTIC EFFECTS OF OLIVE OILS

Extra-virgin olive oil applied topically before or after repeated exposure of mice to ultraviolet B (UVB) delayed the onset of UVB-induced skin tumours. However, with increasing numbers of UVB exposures, differences in the mean number of tumours between UVB control mice and mice pre-treated with olive oil before UVB exposure were lost. In contrast, mice that received olive oil after UVB exposure still showed significantly lower numbers of tumours per mouse than those in the UVB control group throughout the experimental period. Immunohistochemical analysis of DNA damage in the form of cyclobutane pyrimidine dimers (CPD), (6-4) photoproducts and 8-hydroxy-2'-deoxyguanosine (8-OHdG) in samples taken 30 min after a single exposure of UVB, showed no significant difference between UVB-irradiated control mice and the pre-treated UVB group. In the post-UVB group, there were lower levels of 8-OHdG in epidermal nuclei, although the formation of CPD and (6-4) photoproducts did not differ. Significantly, exposure of olive oil to UVB before application abrogated the protective effect on 8-OHdG formation. Taken together, these results indicate that olive oil should be topically applied after UVB exposure to effectively reduce UVB-induced murine skin tumours, since the effective component in oil may be labile to UVB [162]. In X-ray irradiation treatments on micronucleated polychromatic erythrocytes, *Olea europaea* leaves showed a significant anticlastogenic activity both before and after treatment. The activity of these phenolic compounds was based principally on the presence of catechol groups [163]. Concordant with these findings, hydroxytyrosol prevented the increase in ROS induced by UVA irradiation of human melanoma cell line M14. Protective effects were dose-dependent, reaching a maximum at 400 μM. At higher concentrations, hydroxytyrosol acted as a pro-apoptotic stimulus by activating caspase-3 activity. In this cell line, hydroxytyrosol was converted into its methylated derivative, which exhibited a radical scavenging ability comparable to that of hydroxytyrosol [164].

In a separate experimental approach, extracts from olive oil and OMW protected the hydrogen peroxide-induced formation of single strand breaks in nuclear DNA when incubated with Jurkat cells. However, concentrations of olive oil extracts higher than 100 µg/ ml exerted DNA damaging effects by themselves, in the absence of any H_2O_2. Isolated compounds from the extracts were found to exert cytotoxic (oleuropein glucoside), no effect (tyrosol) or protective effects (hydroxytyrosol and caffeic acid). Therefore, the composition of the extracts may drastically change the ultimate effect on DNA damage [165].

Experiments focused on the characterization of proliferative effects have also been designed. Thus, hydroxytyrosol concentrations from 50 to 100 µM caused a complete arrest of human promyelocytic cell line, HL60, in the G0/G1 interphase, with a concomitant decrease of the cell percentages in the S and G2/M phases. Induction of apoptosis was evidenced by flow cytometric analyses, poly(ADP-ribose) polymerase cleavage and caspase-3 activation after 24 h of incubation. Hydroxytyrosol also inhibited proliferation of colon adenocarcinoma cells HT29 and HT29 clone 19A, but the required concentration was higher (750 µM). The apoptotic effect in HL60 cell line required the presence of two ortho-hydroxyl groups on the phenyl ring, since tyrosol did not induce any effect. Hydroxytyrosol-dependent apoptosis was associated with an early release of cytochrome c from mitochondria that preceded caspase-8 activation. Hydroxytyrosol also induced cell death in quiescent and differentiated HL60 cells, as well as in resting and activated peripheral blood lymphocytes, while it did not cause cell death in two colorectal cell lines (HT-29 and Caco2) or freshly isolated human lymphocytes and polymorphonuclear cells. These results suggest that hydroxytyrosol may exert a protective activity against cancer by arresting the cell cycle and inducing apoptosis in tumor cells [166] [167]. In this process, selective activation of transduction pathways such as those of egr1 and c-Jun N-terminal kinase (JNK) were observed and consequently phosphorylation of Bcl-2 [168]. Extract from virgin olive oil enriched in phenolic compounds also inhibited the HL60 cell proliferation in a time- and concentration- dependent manner. At a concentration of 13.5 mg/l, cell growth was completely blocked with an accumulation of cells in the G0/G1 interphase and presence of apoptosis. Two isolated secoiridoids, the dialdehydic forms of elenoic acid linked to hydroxytyrosol and to tyrosol, were shown to possess properties similar to those of whole phenol extract [169].

The investigation of mechanisms involved in hydroxytyrosol-induced growth arrest and apoptosis in human colon carcinoma HT-29 cells has also evidenced prolonged stress of the endoplasmic reticulum (ER), leading to the activation of the unfolded protein response (UPR). Hydroxytyrosol treatment led to overexpression of the pro-apoptotic factor CHOP/GADD153 and persistent activation of the Jun-NH2-terminal kinase/activator protein-1 signalling pathway. Hydroxytyrosol concomitantly modulated the extracellular signal-regulated kinase 1/2 and Akt/PKB pro-survival factors by altering their phosphorylation status as well as inhibiting tumour necrosis factor-alpha-induced nuclear factor-kB activation by inactivating the phosphorylation of nuclear factor inhibitor-kappaB kinase. These findings demonstrate that hydroxytyrosol specifically activates Ser/Thr phosphatase 2A, which plays a key initiating role in various pathways that lead to apoptosis in colon cancer cells [170]. Other pathways could be also involved, as the proposed inhibition of ceramide-induced NF-kB binding activity and apoptosis at micromolar concentrations of caffeic acid in U937 cells through either antioxidant or nonantioxidant mechanisms. Caffeic acid also inhibited protein tyrosine kinase activity, suggesting that this mechanism could form the basis for the inhibition of apoptosis [171]. Collectively, these results support the hypothesis that biophenols play a critical role in the anticancer activity of olive oil.

Chapter 9

Phenolic Compounds Have Anti-Inflammatory Activity

Virgin olive oil with a higher content of phenolic compounds, similar to that of EVOO, showed protective effects in carrageenan and adjuvant arthritis as models of inflammation, and improved the disease associated loss-of-weight in rats [172]. The topical application of oleuropein, tyrosol, hydroxytyrosol or caffeic acid oleuropein (0.5 mg/ear) also reduced the edema induced by either arachidonic acid or 12-O-tetradecanoylphorbol acetate administration in mice, oleuropein being the most effective agent. All compounds strongly inhibited the enzyme myeloperoxidase, indicating a reduction of the neutrophil influx in the inflamed tissues [173]. In a model of systemic inflammation, such as the lipopolysaccharide (LPS)-treated BALB/c mice, olive waste-water at a dose of 125 mg/mouse reduced serum TNF-alpha levels by 95%. In the human monocyte cell line THP-1, olive vegetation water also reduced LPS-induced TNF-alpha production by 50% at a concentration of 0.5 g/l (equivalent to approximately 0.03 g/l simple and polyphenols) [174].

Olive extract strongly inhibited both 12-lipoxygenase (12-LO) and 5-lipoxygenase (5-LO) activities. Hydroxytyrosol inhibited platelet 12-LO activity (IC50, 4.2 μM) and rat polymorphonuclear leukocytes 5-LO activity (IC50, 13 μM) but not cyclooxygenase activity in cell-free conditions. It also inhibited 12-LO activity in intact platelets (IC50, 50 μM) and reduced in a dose-related manner leukotriene B4 production in intact polymorphonuclear leukocytes stimulated by A23187 (IC50, 26 μM) [175] [176]. In rat peritoneal leukocytes stimulated with a calcium ionophore, oleuropein, tyrosol, hydroxytyrosol and caffeic acid also inhibited leukotriene B4 generation by the 5-lipoxygenase activity with a gradient of potency, hydroxytyrosol > oleuropein > caffeic acid > tyrosol. In contrast, none of these compounds caused substantial inhibition of thromboxane generation via the cyclo-oxygenase pathway. Hydroxytyrosol, caffeic acid, oleuropein and tyrosol (decreasing order of potency) also quenched the chemiluminescence signal due to reactive oxygen species generated by phorbol myristate acetate-stimulated rat leukocytes [177]. These results suggest that hydroxytyrosol is a potent specific inhibitor of lipoxygenase activities in cell-free conditions and it could be an effective therapy for a variety of inflammatory processes, including rheumatoid and osteoarthritis. A partial differential outcome was obtained in the macrophage cell line RAW 264.7, where the incubation with tyrosol decreased the O_2^- and H_2O_2 production induced by phorbol esters. These observations were correlated with the impairment of [^{3}H]-arachidonic acid release, cyclooxygenase-2 expression and prostaglandin E_2/leukotriene B_4 synthesis in cultures [178]. The deacetoxy ligstroside aglycone, oleocanthal, has also shown a potency and profile strikingly similar to that of ibuprofen regarding the inhibition of cyclooxygenases [179] [45]. Tyrosol also reduced the *NO release induced by phorbol esters, which was correlated with the impairment of inducible nitric oxide synthase levels. These data point to a modulation of NF-κB activation [178]. A diminished NF-κB activation in mononuclear cells was also observed in volunteer subjects consuming Mediterranean-like diet compared with the same subjects receiving a Western diet [180].

Incubation of J774 murine macrophages with hydroxytyrosol prevented the increase in iNOS and COX-2 mRNA and protein level as well as ROS generation induced by lipopolysaccharide (LPS). In addition, hydroxytyrosol blocked the activation of nuclear factor (NF-κB), signal transducer and activator of transcription-1alpha (STAT-1alpha) and interferon regulatory factor-1 (IRF-1) [181]. On the contrary, oleuropein potentiated the macrophage-mediated response since, in a dose-dependent way, it enhanced nitrite production and iNOS protein expression in LPS-challenged mouse macrophages. This effect was blocked by the iNOS inhibitor L-NAME, indicating increased iNOS activity [182]. Together, these results suggest a potential anti-inflammatory effect with diverse action and mechanisms involved as well as variable cell response for the different phenolic compounds (figure 1A).

Chapter 10

PHENOLIC COMPOUNDS MODULATE ENDOTHELIAL ACTIVITY

Extract from EVOO strongly reduced cell surface expression of intercellular adhesion molecule-1 (ICAM-1) and vascular cell adhesion molecule-1 (VCAM-1) at physiologically relevant concentrations (IC50 < 1 μM), linked to a reduction in mRNA levels in human umbilical vascular endothelial cells. Oleuropein aglycone and hydroxytyrosol were mainly responsible for these effects. Homovanillyl alcohol inhibited cell surface expression of all the adhesion molecules, whereas the effect on mRNA expression was weaker [183]. Oleuropein and hydroxytyrosol also reduced the monocytoid cell adhesion to lipopolysaccharide or cytokine- stimulated human umbilical vein endothelial cells. Reporter gene assays with deletional VCAM-1 promoter constructs indicated the relevance of nuclear factor-κB, activator protein-1, and possibly GATA binding sites in mediating VCAM-1 transcriptional inhibition [184]. These results were also observed when human umbilical vein endothelial cells were exposed to the cytotoxic effects of hydrogen peroxide and oxidized LDL, since oleuropein, tyrosol and hydroxytyrosol induced a significant reduction in the secretion of ICAM-1 and VCAM-1. On the other hand, no influence on nitric oxide production in this experimental setting was reported [133]. Using the latter experimental approach, it has been identified that LDL isolated from subjects consuming a Mediterranean or a low-fat high-carbohydrate diet induced a lower expression of VCAM-1 and E-selectin when compared with those coming from the saturated fat-enriched diet [141]. These results suggest that both the Mediterranean and high carbohydrate diets may decrease the pro-inflammatory environment induced by modified LDL on endothelial cells (Figure 1A).

Resting blood pressure was significantly lower at the end of the EVOO diet compared with the sunflower oil diet. Blood pressure responses during sympathetic stimulation with the cold pressor test and isometric exercise were similar. Daily drug dosage was significantly reduced during the extra virgin olive oil diet but not the polyunsaturated diet. A slight reduction in saturated fat intake, along with the use of extra-virgin olive oil, markedly lowers daily antihypertensive dosage requirement [185].

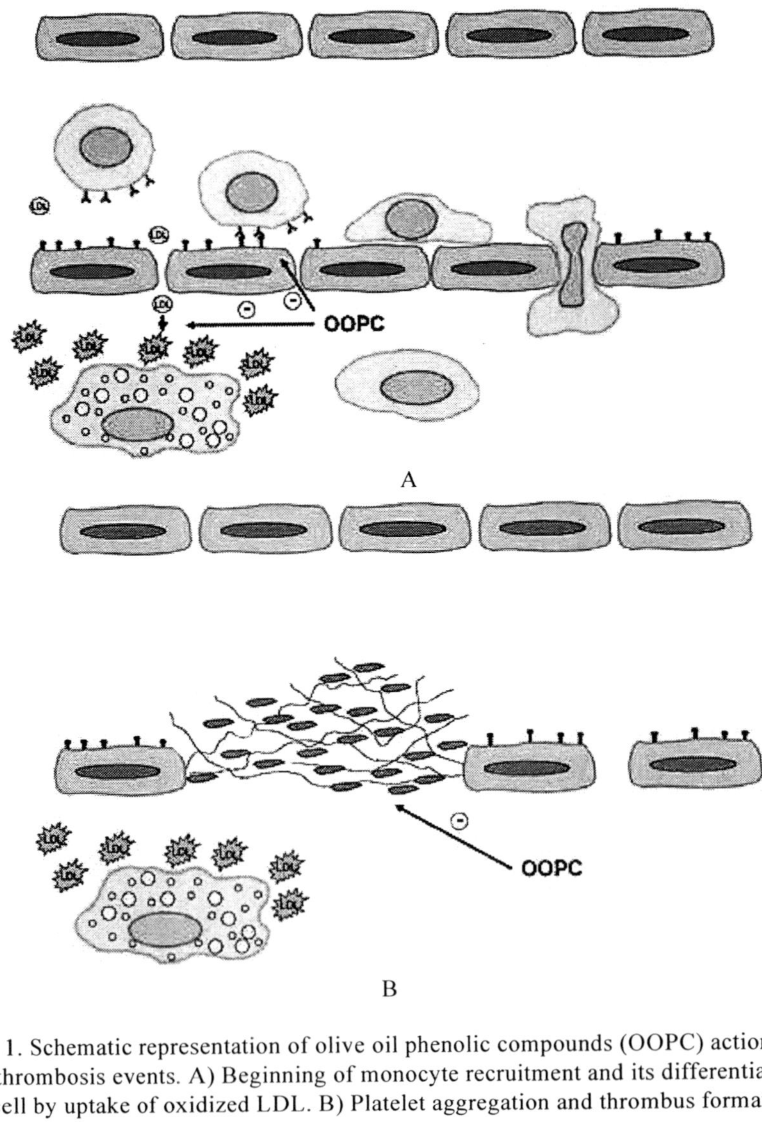

A

B

Figure 1. Schematic representation of olive oil phenolic compounds (OOPC) actions on the atherothrombosis events. A) Beginning of monocyte recruitment and its differentiation into foam cell by uptake of oxidized LDL. B) Platelet aggregation and thrombus formation.

In aortic ring preparations, the contraction in response to phenylephrine was significantly decreased in spontaneously hypertensive rats fed with virgin olive oil- in comparison with high-oleic-acid sunflower oil- containing diets. Equally, the relaxant responses to acetylcholine were significantly enhanced in spontaneously hypertensive rats fed the virgin olive oil-enriched diet, suggesting that the chronic feeding of olive oil diet was able to attenuate the vascular response of rat aorta.

Thus, other components of olive oil, not only oleic acid, may help to explain the vascular protective effect of olive oil consumption [186]. In *ex vivo* aortic rings, hydroxytyrosol (10 μM) efficiently protected the aorta against the cumene hydroperoxide-induced impairment of the NO mediated relaxation. Oleuropein, tyrosol and homovanillic alcohol did not show any protection [114].

Chapter 11

PHENOLIC COMPOUNDS MODULATE PLATELET AGGREGATION

Preincubation of platelet rich plasma (PRP) with hydroxytyrosol for at least 10 min resulted in maximal inhibition of platelet aggregation induced by ADP or collagen, and of thromboxane B2 (TxB2) production by collagen or thrombin-stimulated PRP. The hydroxytyrosol concentrations resulting in 50% inhibition of ADP or collagen-induced PRP aggregations were 23 and 67 μM, respectively. A higher concentration of 400 μM was required to completely inhibit collagen-induced PRP aggregation. Likewise, TxB2 production by collagen- or thrombin-stimulated PRP was inhibited by over 80 percent. At the same hydroxytyrosol concentration, the accumulation of TxB2 and 12-HETE in serum was reduced by over 90 and 50 percent, respectively. The effects on PRP aggregation of oleuropein were found to be much less active. On the other hand, a phenol-enriched extract obtained from aqueous waste of olive oil showed rather potent activities [187]. An anti-aggregating platelet action was also found after the administration of 25 mg of hydroxytyrosol-rich phenolic extract the first day, and 12.5 mg/day the following 3 days to five type-I diabetic patients that resulted in a 46% decrease in the serum TxB2 production after blood clotting at the end of the experiment [188]. Two isochromans [1-(3'-methoxy-4'-hydroxy-phenyl)-6,7-dihydroxy-isochroman and 1-phenyl-6,7-dihydroxy-isochroman], recently discovered in olive oil, were also effective as free radical scavengers, inhibited platelet aggregation and thromboxane release evoked by agonists that induce ROS-mediated platelet activation, including sodium arachidonate and collagen but not ADP. Release of tritiated arachidonic acid from platelets was also impaired [189]. Overall, components of the phenolic fraction of olive oil can inhibit platelet function and eicosanoid formation *in vitro* (Figure 1B).

Chapter 12

Role of Olive Oils in Atherosclerosis

Using either EVOOs at 10% (w/w) with different oleic acid content or sunflower oils enriched in this fatty acid, we have shown that the presence of oleic acid *per se* was unable to explain the effects of extra virgin olive oil upon atherosclerosis apoE- deficient mice [190]. In this animal model, the inclusion of olive oil at 10% (w/w) in the diet was more effective in preventing atherosclerosis development than when this component was present at 20%, suggesting the existence of a dose window in which olive oil could exert its beneficial effects. Likewise, using diets containing EVOOs with or without cholesterol, we have significantly demonstrated that in the absence of dietary cholesterol, olive oil-containing diets are more effective in delaying atherosclerosis progression [191]. To gain a better understanding of the olive oil components involved in the anti-atherosclerosis effect, new experiments have been carried out by our group. Surprisingly, administration of hydroxytyrosol, at a dose of 10 mg/kg/day to ApoE- deficient male mice on a standard chow diet of low cholesterol content, increased atherosclerotic lesions mainly associated with the degree of monocyte activation and an increase in total cholesterol. These results indicate that phenolic-enriched products, out of the original matrix, could actually be harmful. Our results suggest that the formulation of possible functional foods should approximate as much as possible the natural environment in which the active molecules are found [192]. In this sense, when a mixture of biophenols containing cathechin, caffeic acid and resveratrol at 40 or 160 mg/kg /day dose was administered to ApoE- deficient mice, the presence of atherosclerosis was decreased by modulating inflammatory cells recruitment and expression of pro-inflammatory chemokines in the vascular wall [193]. These results provide support to our data in the sense that the complex mixture of phenolic compound of extra virgin olive oil would be necessary. In a different animal model, administration of hydroxytyrosol at 4 mg/kg body weight in presence of an atherogenic diet high in saturated fat and cholesterol reduced the size of atherosclerotic lesions in rabbits compared with the animals receiving this diet without hydroxytyrosol [194]. Collectively, these works suggest that the regime of hydroxytyrosol administration could be important as well as the model of assay. Potential interaction with dietary cholesterol or with other phenolic compounds needs to be further considered as well as the response in different animal models with special adaptations to lipid metabolism.

Chapter 13

Conclusion

Extra virgin olive oil is a non-oxidative dietary component with a whole plethora of biological actions (Table 2), with a particularly remarkable effect on the attenuation of inflammatory processes. These two actions could explain its beneficial effects on disease risk since oxidative and inflammatory stresses appear to be underlying factors in the aetiology of many age-related diseases in man. The antioxidant effects of olive oil are probably due to a combination of its high oleic acid content and its content of a variety of plant antioxidants, particularly phenolic compounds. In fact, the phenolic compounds in virgin olive oil explain some of the protective effects found in epidemiological studies [195] and the biological properties of olive oil [196]. Adoption of a Mediterranean diet could confer health benefits in high-risk populations [197].

Numerous bioactive compounds, in addition, appear to have beneficial health effects. Thus, research on the pharmacological properties of the minor components of olive oil, such as biophenols, is very active and might lead to the formulation of functional foods and nutraceuticals. Much scientific research, either in its basic or interventional aspects, still needs to be conducted before we can begin to make science-based dietary recommendations [24]. Despite this, epidemiologic data, together with experimental pathology and cell biology, supports the recommendation that optimal nutrition should contain biophenols in amounts that may be better described as a "Recommended Optimal Intake" [198]. Furthermore, recent evidence recommends their consumption in foods that are as fresh as possible.

Table 2. Summary of biological actions for the different phenolic compounds

	Anti-oxidant	Anti-Microbial and antiviral	Anti LDL oxidation	Defence system	Anti-proliferative and apoptotic	Anti-inflammatory
Hydroxytyrosol	X	X	X	X	X	X
Tyrosol			X	X		X
Homovanillic alcohol	X		X			
Caffeic acid	X				X	X
Rutin, quercetin, luteolin			X			
Oleuropein	X	X	X			X
Oleuropein glycoside			X	X		
Ligstroside derivates		X	X			

ACKNOWLEDGMENTS

The research of this group was supported by Grants CICYT- FEDER (SAF2004-08173-C03-02 and SAF2007-60173) and FISS (01/0202), *Redes DGA (*B-69*), FISS de Investigación Cooperativa* (G03/140 and C03/01) and by *Fundación Española del Corazón.* M.A.N. was recipient of a FEGA-FEOGA fellowship.

REFERENCES

[1] Kromhout D. Epidemiology of cardiovascular diseases in Europe. *Public Health Nutr*. 2001;4:441-57.

[2] Simopoulos AP. The Mediterranean diets: What is so special about the diet of Greece? The scientific evidence. *J. Nutr*. 2001;131:3065S-73S.

[3] Kromhout D. Diet and cardiovascular diseases. *J. Nutr. Health Aging* 2001;5:144-9.

[4] Trichopoulou A, Costacou T, Bamia C, Trichopoulos D. Adherence to a Mediterranean Diet and Survival in a Greek Population. *N. Engl. J. Med.* 2003;348:2599-2608.

[5] Knoops KTB, de Groot LCPGM, Kromhout D, Perrin A-E, Moreiras-Varela O, Menotti A, et al. Mediterranean Diet, Lifestyle Factors, and 10-Year Mortality in Elderly European Men and Women: The HALE Project. *JAMA* 2004;292:1433-1439.

[6] Owen RW, Giacosa A, Hull WE, Haubner R, Wurtele G, Spiegelhalder B, et al. Olive-oil consumption and health: the possible role of antioxidants. *Lancet Oncol*. 2000;1:107-12.

[7] Owen RW, Haubner R, Wurtele G, Hull E, Spiegelhalder B, Bartsch H. Olives and olive oil in cancer prevention. *Eur. J. Cancer Prev*. 2004;13:319-26.

[8] Perona JS, Cabello-Moruno R, Ruiz-Gutierrez V. The role of virgin olive oil components in the modulation of endothelial function. *J. Nutr. Biochem*. 2006;17:429-45.

[9] Tovar MJ, Motilva MJ, Romero MP. Changes in the phenolic composition of virgin olive oil from young trees (Olea europaea L. cv. Arbequina) grown under linear irrigation strategies. *J. Agric. Food Chem*. 2001;49:5502-8.

[10] Brenes M, Romero C, Garcia A, Hidalgo FJ, Ruiz-Mendez MV. Phenolic compounds in olive oils intended for refining: formation of 4-ethylphenol during olive paste storage. *J. Agric. Food Chem.* 2004;52:8177-81.

[11] Romero C, Brenes M, Yousfi K, Garcia P, Garcia A, Garrido A. Effect of cultivar and processing method on the contents of polyphenols in table olives. *J. Agric. Food Chem.* 2004;52:479-84.

[12] Rios JJ, Gil MJ, Gutierrez-Rosales F. Solid-phase extraction gas chromatography-ion trap-mass spectrometry qualitative method for evaluation of phenolic compounds in virgin olive oil and structural confirmation of oleuropein and ligstroside aglycons and their oxidation products. *J. Chromatogr.* A 2005;1093:167-76.

[13] Brenes M, Garcia A, Garcia P, Rios JJ, Garrido A. Phenolic compounds in Spanish olive oils. *J. Agric. Food Chem.* 1999;47:3535-40.

[14] Bonoli M, Montanucci M, Gallina Toschi T, Lercker G. Fast separation and determination of tyrosol, hydroxytyrosol and other phenolic compounds in extra-virgin olive oil by capillary zone electrophoresis with ultraviolet-diode array detection. *J. Chromatogr.* A 2003;1011:163-72.

[15] Bonoli M, Bendini A, Cerretani L, Lercker G, Toschi TG. Qualitative and semiquantitative analysis of phenolic compounds in extra virgin olive oils as a function of the ripening degree of olive fruits by different analytical techniques. *J. Agric. Food Chem.* 2004;52:7026-32.

[16] Esti M, Cinquanta L, La Notte E. Phenolic Compounds in Different Olive Varieties. *J. Agric. Food Chem.* 1998;46:32-35.

[17] Caruso D, Colombo R, Patelli R, Giavarini F, Galli G. Rapid evaluation of phenolic component profile and analysis of oleuropein aglycon in olive oil by atmospheric pressure chemical ionization-mass spectrometry (APCI-MS). *J. Agric. Food Chem.* 2000;48:1182-5.

[18] Gomez-Alonso S, Salvador MD, Fregapane G. Phenolic compounds profile of cornicabra virgin olive oil. *J. Agric. Food Chem.* 2002;50:6812-7.

[19] Carrasco-Pancorbo A, Arraez-Roman D, Segura-Carretero A, Fernandez-Gutierrez A. Capillary electrophoresis-electrospray ionization-mass spectrometry method to determine the phenolic fraction of extra-virgin olive oil. *Electrophoresis* 2006;27:2182-96.

[20] Christophoridou S, Dais P, Tseng LH, Spraul M. Separation and identification of phenolic compounds in olive oil by coupling high-performance liquid chromatography with postcolumn solid-phase extraction to nuclear magnetic resonance spectroscopy (LC-SPE-NMR). *J. Agric. Food Chem.* 2005;53:4667-79.

[21] Christophoridou S, Dais P. Novel approach to the detection and quantification of phenolic compounds in olive oil based on 31P nuclear magnetic resonance spectroscopy. *J. Agric. Food Chem.* 2006;54:656-64.

[22] Dais P, Spyros A, Christophoridou S, Hatzakis E, Fragaki G, Agiomyrgianaki A, et al. Comparison of Analytical Methodologies Based on (1)H and (31)P NMR Spectroscopy with Conventional Methods of Analysis for the Determination of Some Olive Oil Constituents. *J. Agric. Food Chem.* 2007;55:577-584.

[23] Carrasco-Pancorbo A, Cerretani L, Bendini A, Segura-Carretero A, Gallina-Toschi T, Fernandez-Gutierrez A. Analytical determination of polyphenols in olive oils. *J. Sep. Sci.* 2005;28:837-58.

[24] Covas MI, Ruiz-Gutierrez V, De la Torre R, Kafatos A, Lamuela-Raventos RM, Osada J, et al. Minor components of olive oil: evidence to date of health benefits in humans. *Nutrition Reviews* 2006;64:S20-S30.

[25] Ranalli A, Ferrante ML, De Mattia G, Costantini N. Analytical evaluation of virgin olive oil of first and second extraction. *J. Agric. Food Chem.* 1999;47:417-24.

[26] Owen RW, Mier W, Giacosa A, Hull WE, Spiegelhalder B, Bartsch H. Phenolic compounds and squalene in olive oils: the concentration and antioxidant potential of total phenols, simple phenols, secoiridoids, lignansand squalene. *Food Chem. Toxicol.* 2000;38:647-59.

[27] DellaGreca M, Previtera L, Temussi F, Zarrelli A. Low-molecular-weight components of olive oil mill waste-waters. *Phytochem. Anal.* 2004;15:184-8.

[28] Tripoli E, Giammanco M, Tabacchi G, Di Majo D, Giammanco S, La Guardia M. The phenolic compound of olive oil: structure, biological activity and beneficial effects on human health. *Nutrition Research Reviews* 2005;18:98-112.

[29] Bouaziz M, Chamkha M, Sayadi S. Comparative study on phenolic content and antioxidant activity during maturation of the olive cultivar Chemlali from Tunisia. *J. Agric. Food Chem.* 2004;52:5476-81.

[30] Pereira JA, Pereira AP, Ferreira IC, Valentao P, Andrade PB, Seabra R, et al. Table olives from Portugal: phenolic compounds, antioxidant potential, and antimicrobial activity. *J. Agric. Food Chem.* 2006;54:8425-31.

[31] Romero C, Brenes M, Garcia P, Garrido A. Hydroxytyrosol 4-beta-D-glucoside, an important phenolic compound in olive fruits and derived products. *J. Agric. Food Chem.* 2002;50:3835-9.

[32] Romero C, Brenes M, Garcia P, Garcia A, Garrido A. Polyphenol changes during fermentation of naturally black olives. *J. Agric. Food Chem.* 2004;52:1973-9.

[33] Dourtoglou VG, Mamalos A, Makris DP, Kefalas P. Storage of olives (Olea europaea L.) under CO2 atmosphere: liquid chromatography-mass spectrometry characterization of indices related to changes in polyphenolic metabolism. *J. Agric. Food Chem.* 2006;54:2211-7.

[34] Owen RW, Haubner R, Mier W, Giacosa A, Hull WE, Spiegelhalder B, et al. Isolation, structure elucidation and antioxidant potential of the major phenolic and flavonoid compounds in brined olive drupes. *Food Chem. Toxicol.* 2003;41:703-17.

[35] Blekas G, Vassilakis C, Harizanis C, Tsimidou M, Boskou DG. Biophenols in table olives. *J. Agric. Food Chem.* 2002;50:3688-92.

[36] Ben Miled DD, Smaoui A, Zarrouk M, Cherif A. Do extraction procedures affect olive oil quality and stability? *Biochem. Soc. Trans.* 2000;28:929-33.

[37] Brenes M, Garcia A, Garcia P, Garrido A. Acid hydrolysis of secoiridoid aglycons during storage of virgin olive oil. *J. Agric. Food Chem.* 2001;49:5609-14.

[38] Gomez Caravaca AM, Carrasco Pancorbo A, Canabate Diaz B, Segura Carretero A, Fernandez Gutierrez A. Electrophoretic identification and quantitation of compounds in the polyphenolic fraction of extra-virgin olive oil. *Electrophoresis* 2005;26:3538-51.

[39] Arbonés-Mainar JM, Navarro MA, Carnicer R, Guillén N, Surra JC, Acín S, et al. Accelerated atherosclerosis in apolipoprotein E-deficient mice fed Western diets containing palm oil compared with extra virgin olive oils: A role for small, dense High-Density Lipoproteins. *Atherosclerosis.* 2006:http://doi:10.1016/j.atherosclerosis.

[40] 2006.11.010.

[41] Gomez-Rico A, Salvador MD, La Greca M, Fregapane G. Phenolic and volatile compounds of extra virgin olive oil (Olea europaea L. Cv. Cornicabra) with regard to fruit ripening and irrigation management. *J. Agric. Food Chem.* 2006;54:7130-6.

[42] Gutierrez F, Fernandez JL. Determinant parameters and components in the storage of virgin olive oil. Prediction of storage time beyond which the oil is no longer of "extra" quality. *J. Agric. Food Chem.* 2002;50:571-7.

[43] Lavelli V, Bondesan L. Secoiridoids, tocopherols, and antioxidant activity of monovarietal extra virgin olive oils extracted from destoned fruits. *J. Agric. Food Chem.* 2005;53:1102-7.

[44] Gutierrez-Rosales F, Rios JJ, Gomez-Rey ML. Main polyphenols in the bitter taste of virgin olive oil. Structural confirmation by on-line high-performance liquid chromatography electrospray ionization mass spectrometry. *J. Agric. Food Chem.* 2003;51:6021-5.

[45] Garcia JM, Yousfi K, Mateos R, Olmo M, Cert A. Reduction of oil bitterness by heating of olive (Olea europaea) fruits. *J. Agric. Food Chem.* 2001;49:4231-5.

[46] Impellizzeri J, Lin J. A simple high-performance liquid chromatography method for the determination of throat-burning oleocanthal with probated antiinflammatory activity in extra virgin olive oils. *J. Agric. Food Chem.* 2006;54:3204-8.

[47] Andrewes P, Busch JL, de Joode T, Groenewegen A, Alexandre H. Sensory properties of virgin olive oil polyphenols: identification of deacetoxy-ligstroside aglycon as a key contributor to pungency. *J. Agric. Food Chem.* 2003;51:1415-20.

[48] Visioli F, Galli C. Biological properties of olive oil phytochemicals. *Crit. Rev. Food Sci. Nutr.* 2002;42:209-21.

[49] Visioli F, Poli A, Gall C. Antioxidant and other biological activities of phenols from olives and olive oil. *Med. Res. Rev.* 2002;22:65-75.

[50] Valavanidis A, Nisiotou C, Papageorgiou Y, Kremli I, Satravelas N, Zinieris N, et al. Comparison of the radical scavenging potential of polar and lipidic fractions of olive oil and other vegetable oils under normal conditions and after thermal treatment. *J. Agric. Food Chem.* 2004;52:2358-65.

[51] Pellegrini N, Visioli F, Buratti S, Brighenti F. Direct analysis of total antioxidant activity of olive oil and studies on the influence of heating. *J. Agric. Food Chem.* 2001;49:2532-8.

[52] Del Carlo M, Sacchetti G, Di Mattia C, Compagnone D, Mastrocola D, Liberatore L, et al. Contribution of the phenolic fraction to the antioxidant activity and oxidative stability of olive oil. *J. Agric. Food Chem.* 2004;52:4072-9.

[53] Gorinstein S, Martin-Belloso O, Katrich E, Lojek A, Ciz M, Gligelmo-Miguel N, et al. Comparison of the contents of the main biochemical compounds and the antioxidant activity of some Spanish olive oils as determined by four different radical scavenging tests. *J. Nutr. Biochem.* 2003;14:154-9.

[54] Mateos R, Trujillo M, Perez-Camino MC, Moreda W, Cert A. Relationships between oxidative stability, triacylglycerol composition, and antioxidant content in olive oil matrices. *J. Agric. Food Chem.* 2005;53:5766-71.

[55] Mateos R, Dominguez MM, Espartero JL, Cert A. Antioxidant effect of phenolic compounds, alpha-tocopherol, and other minor components in virgin olive oil. *J. Agric. Food Chem.* 2003;51:7170-5.

[56] Bouaziz M, Grayer RJ, Simmonds MS, Damak M, Sayadi S. Identification and antioxidant potential of flavonoids and low molecular weight phenols in olive cultivar chemlali growing in Tunisia. *J. Agric. Food Chem.* 2005;53:236-41.

[57] Carrasco-Pancorbo A, Cerretani L, Bendini A, Segura-Carretero A, Del Carlo M, Gallina-Toschi T, et al. Evaluation of the antioxidant capacity of individual phenolic compounds in virgin olive oil. *J. Agric. Food Chem.* 2005;53:8918-25.

[58] Franconi F, Coinu R, Carta S, Urgeghe PP, Ieri F, Mulinacci N, et al. Antioxidant effect of two virgin olive oils depends on the concentration and composition of minor polar compounds. *J. Agric. Food Chem.* 2006;54:3121-5.

[59] Lavelli V. Comparison of the antioxidant activities of extra virgin olive oils. J Agric Food Chem 2002;50:7704-8.

[60] Perez-Bonilla M, Salido S, van Beek TA, Linares-Palomino PJ, Altarejos J, Nogueras M, et al. Isolation and identification of radical scavengers in olive tree (Olea europaea) wood. *J. Chromatogr.* A 2006;1112:311-8.

[61] Tuck KL, Hayball PJ. Major phenolic compounds in olive oil: metabolism and health effects. *J. Nutr. Biochem.* 2002;13:636-644.

[62] Tuck KL, Hayball PJ, Stupans I. Structural characterization of the metabolites of hydroxytyrosol, the principal phenolic component in olive oil, in rats. *J. Agric. Food Chem.* 2002;50:2404-9.

[63] Nenadis N, Wang LF, Tsimidou MZ, Zhang HY. Radical scavenging potential of phenolic compounds encountered in O. europaea products as indicated by calculation of bond dissociation enthalpy and ionization potential values. *J. Agric. Food Chem.* 2005;53:295-9.

[64] Roche M, Dufour C, Mora N, Dangles O. Antioxidant activity of olive phenols: mechanistic investigation and characterization of oxidation products by mass spectrometry. *Org. Biomol. Chem.* 2005;3:423-30.

[65] Paiva-Martins F, Gordon MH, Gameiro P. Activity and location of olive oil phenolic antioxidants in liposomes. *Chem. Phys. Lipids* 2003;124:23-36.

[66] Medina I, Tombo I, Satue-Gracia MT, German JB, Frankel EN. Effects of natural phenolic compounds on the antioxidant activity of lactoferrin in liposomes and oil-in-water emulsions. *J. Agric. Food Chem.* 2002;50:2392-9.

[67] Morello JR, Vuorela S, Romero MP, Motilva MJ, Heinonen M. Antioxidant activity of olive pulp and olive oil phenolic compounds of the arbequina cultivar. *J. Agric. Food Chem.* 2005;53:2002-8.

[68] Stupans I, Kirlich A, Tuck KL, Hayball PJ. Comparison of radical scavenging effect, inhibition of microsomal oxygen free radical generation, and serum lipoprotein oxidation of several natural antioxidants. *J. Agric. Food Chem.* 2002;50:2464-9.

[69] Paiva-Martins F, Gordon MH. Interactions of ferric ions with olive oil phenolic compounds. *J. Agric. Food Chem.* 2005;53:2704-9.

[70] Pazos M, Alonso A, Fernandez-Bolanos J, Torres JL, Medina I. Physicochemical properties of natural phenolics from grapes and olive oil byproducts and their antioxidant activity in frozen horse mackerel fillets. *J. Agric. Food Chem.* 2006;54:366-73.

[71] Medina I, Satue-Gracia MT, German JB, Frankel EN. Comparison of natural polyphenol antioxidants from extra virgin olive oil with synthetic antioxidants in tuna lipids during thermal oxidation. *J. Agric. Food Chem.* 1999;47:4873-9.

[72] Pazos M, Sanchez L, Medina I. Alpha-tocopherol oxidation in fish muscle during chilling and frozen storage. *J. Agric. Food Chem.* 2005;53:4000-5.

[73] Pazos M, Lois S, Torres JL, Medina I. Inhibition of hemoglobin- and iron-promoted oxidation in fish microsomes by natural phenolics. *J. Agric. Food Chem.* 2006;54:4417-23.

[74] Brenes M, Garcia A, Dobarganes MC, Velasco J, Romero C. Influence of thermal treatments simulating cooking processes on the polyphenol content in virgin olive oil. *J. Agric. Food Chem.* 2002;50:5962-7.

[75] Gomez-Alonso S, Fregapane G, Salvador MD, Gordon MH. Changes in phenolic composition and antioxidant activity of virgin olive oil during frying. *J. Agric. Food Chem.* 2003;51:667-72.

[76] Andrikopoulos NK, Dedoussis GV, Falirea A, Kalogeropoulos N, Hatzinikola HS. Deterioration of natural antioxidant species of vegetable edible oils during the domestic deep-frying and pan-frying of potatoes. *Int. J. Food Sci. Nutr.* 2002;53:351-63.

[77] Cowan MM. Plant products as antimicrobial agents. *Clin. Microbiol. Rev.* 1999;12:564-82.

[78] Bisignano G, Tomaino A, Lo Cascio R, Crisafi G, Uccella N, Saija A. On the in-vitro antimicrobial activity of oleuropein and hydroxytyrosol. *J. Pharm. Pharmacol.* 1999;51:971-4.

[79] Furneri PM, Piperno A, Sajia A, Bisignano G. Antimycoplasmal activity of hydroxytyrosol. *Antimicrob. Agents Chemother* 2004;48:4892-4.

[80] Romero C, Medina E, Vargas J, Brenes M, Castro AD. In Vitro Activity of Olive Oil Polyphenols against Helicobacter pylori. *J. Agric. Food Chem.* 2007;55:680-6.

[81] Medina E, de Castro A, Romero C, Brenes M. Comparison of the concentrations of phenolic compounds in olive oils and other plant oils: correlation with antimicrobial activity. *J. Agric. Food Chem.* 2006;54:4954-61.

[82] Lee-Huang S, Huang PL, Zhang D, Lee JW, Bao J, Sun Y, et al. Discovery of small-molecule HIV-1 fusion and integrase inhibitors oleuropein and hydroxytyrosol: Part I. Fusion inhibition. *Biochem. Biophys. Res. Commun.* 2007.

[83] Lee-Huang S, Huang PL, Zhang D, Lee JW, Bao J, Sun Y, et al. Discovery of small-molecule HIV-1 fusion and integrase inhibitors oleuropein and hydroxytyrosol: II. Integrase inhibition. *Biochem. Biophys. Res. Commun.* 2007.

[84] Ruiz-Gutierrez V, Juan ME, Cert A, Planas JM. Determination of hydroxytyrosol in plasma by HPLC. *Anal. Chem.* 2000;72:4458-61.

[85] Tan HW, Tuck KL, Stupans I, Hayball PJ. Simultaneous determination of oleuropein and hydroxytyrosol in rat plasma using liquid chromatography with fluorescence detection. *J. Chromatogr. B Analyt. Technol. Biomed. Life Sci.* 2003;785:187-91.

[86] Bazoti FN, Gikas E, Puel C, Coxam V, Tsarbopoulos A. Development of a sensitive and specific solid phase extraction--gas chromatography-tandem mass spectrometry method for the determination of elenolic acid, hydroxytyrosol, and tyrosol in rat urine. *J. Agric. Food Chem.* 2005;53:6213-21.

[87] Corona G, Tzounis X, Assunta Dessi M, Deiana M, Debnam ES, Visioli F, et al. The fate of olive oil polyphenols in the gastrointestinal tract: implications of gastric and colonic microflora-dependent biotransformation. *Free Radic. Res.* 2006;40:647-58.

[88] Manna C, Galletti P, Maisto G, Cucciolla V, D'Angelo S, Zappia V. Transport mechanism and metabolism of olive oil hydroxytyrosol in Caco-2 cells. *FEBS Lett* 2000;470:341-4.

[89] Edgecombe SC, Stretch GL, Hayball PJ. Oleuropein, an antioxidant polyphenol from olive oil, is poorly absorbed from isolated perfused rat intestine. *J. Nutr.* 2000;130:2996-3002.

[90] Christian MS, Sharper VA, Hoberman AM, Seng JE, Fu L, Covell D, et al. The toxicity profile of hydrolyzed aqueous olive pulp extract. *Drug Chem. Toxicol.* 2004;27:309-30.

[91] Soni MG, Burdock GA, Christian MS, Bitler CM, Crea R. Safety assessment of aqueous olive pulp extract as an antioxidant or antimicrobial agent in foods. *Food Chem. Toxicol.* 2006;44:903-15.

[92] Visioli F, Caruso D, Plasmati E, Patelli R, Mulinacci N, Romani A, et al. Hydroxytyrosol, as a component of olive mill waste water, is dose-dependently absorbed and increases the antioxidant capacity of rat plasma. *Free Radic. Res.* 2001;34:301-5.

[93] Tuck KL, Freeman MP, Hayball PJ, Stretch GL, Stupans I. The in vivo fate of hydroxytyrosol and tyrosol, antioxidant phenolic constituents of olive oil, after intravenous and oral dosing of labeled compounds to rats. *J. Nutr.* 2001;131:1993-6.

[94] Bonanome A, Pagnan A, Caruso D, Toia A, Xamin A, Fedeli E, et al. Evidence of postprandial absorption of olive oil phenols in humans. *Nutr. Metab. Cardiovasc. Dis.* 2000;10:111-20.

[95] Masella R, Giovannini C, Vari R, Di Benedetto R, Coni E, Volpe R, et al. Effects of dietary virgin olive oil phenols on low density lipoprotein oxidation in hyperlipidemic patients. *Lipids* 2001;36:1195-202.

[96] Miro-Casas E, Covas MI, Fito M, Farre-Albadalejo M, Marrugat J, de la Torre R. Tyrosol and hydroxytyrosol are absorbed from moderate and sustained doses of virgin olive oil in humans. *Eur. J. Clin. Nutr.* 2003;57:186-90.

[97] Vissers MN, Zock PL, Roodenburg AJ, Leenen R, Katan MB. Olive oil phenols are absorbed in humans. *J. Nutr.* 2002;132:409-417.

[98] Weinbrenner T, Fito M, Farre Albaladejo M, Saez GT, Rijken P, Tormos C, et al. Bioavailability of phenolic compounds from olive oil and oxidative/antioxidant status at postprandial state in healthy humans. *Drugs Exp. Clin. Res.* 2004;30:207-12.

[99] Covas MI, de la Torre K, Farre-Albaladejo M, Kaikkonen J, Fito M, Lopez-Sabater C, et al. Postprandial LDL phenolic content and LDL oxidation are modulated by olive oil phenolic compounds in humans. *Free Radic. Biol. Med.* 2006;40:608-16.

[100] Scalbert A, Deprez S, Mila I, Albrecht AM, Huneau JF, Rabot S. Proanthocyanidins and human health: systemic effects and local effects in the gut. *Biofactors* 2000;13:115-20.

[101] Piskula MK, Terao J. Accumulation of (-)-epicatechin metabolites in rat plasma after oral administration and distribution of conjugation enzymes in rat tissues. *J. Nutr.* 1998;128:1172-8.

[102] Mateos R, Goya L, Bravo L. Metabolism of the olive oil phenols hydroxytyrosol, tyrosol, and hydroxytyrosyl acetate by human hepatoma HepG2 cells. *J. Agric. Food Chem.* 2005;53:9897-905.

[103] Del Boccio P, Di Deo A, De Curtis A, Celli N, Iacoviello L, Rotilio D. Liquid chromatography-tandem mass spectrometry analysis of oleuropein and its metabolite hydroxytyrosol in rat plasma and urine after oral administration. *J. Chromatogr. B Analyt. Technol. Biomed. Life Sci.* 2003;785:47-56.

[104] Visioli F, Galli C, Grande S, Colonnelli K, Patelli C, Galli G, et al. Hydroxytyrosol excretion differs between rats and humans and depends on the vehicle of administration. *J. Nutr.* 2003;133:2612-5.

[105] de la Torre-Carbot K, Jauregui O, Castellote AI, Lamuela-Raventos RM, Covas MI, Casals I, et al. Rapid high-performance liquid chromatography-electrospray ionization tandem mass spectrometry method for qualitative and quantitative analysis of virgin olive oil phenolic metabolites in human low-density lipoproteins. *J. Chromatogr.* A 2006;1116:69-75.

[106] D'Angelo S, Manna C, Migliardi V, Mazzoni O, Morrica P, Capasso G, et al. Pharmacokinetics and metabolism of hydroxytyrosol, a natural antioxidant from olive oil. *Drug Metab. Dispos.* 2001;29:1492-8.

[107] Caruso D, Visioli F, Patelli R, Galli C, Galli G. Urinary excretion of olive oil phenols and their metabolites in humans. *Metabolism* 2001;50:1426-8.

[108] Visioli F, Galli C, Bornet F, Mattei A, Patelli R, Galli G, et al. Olive oil phenolics are dose-dependently absorbed in humans. *FEBS Lett.* 2000;468:159-60.

[109] Miro-Casas E, Farre Albaladejo M, Covas MI, Rodriguez JO, Menoyo Colomer E, Lamuela Raventos RM, et al. Capillary gas chromatography-mass spectrometry quantitative determination of hydroxytyrosol and tyrosol in human urine after olive oil intake. *Anal. Biochem.* 2001;294:63-72.

[110] Miro-Casas E, Covas MI, Farre M, Fito M, Ortuno J, Weinbrenner T, et al. Hydroxytyrosol disposition in humans. *Clin. Chem.* 2003;49:945-52.

[111] Vissers MN, Zock PL, Katan MB. Bioavailability and antioxidant effects of olive oil phenols in humans: a review. *Eur. J. Clin. Nutr.* 2004;58:955-65.

[112] Manach C, Mazur A, Scalbert A. Polyphenols and prevention of cardiovascular diseases. *Curr. Opin. Lipidol.* 2005;16:77-84.

[113] O'Dowd Y, Driss F, Dang PM, Elbim C, Gougerot-Pocidalo MA, Pasquier C, et al. Antioxidant effect of hydroxytyrosol, a polyphenol from olive oil: scavenging of hydrogen peroxide but not superoxide anion produced by human neutrophils. *Biochem. Pharmacol.* 2004;68:2003-8.

[114] Visioli F, Bellomo G, Galli C. Free radical-scavenging properties of olive oil polyphenols. *Biochem. Biophys. Res. Commun.* 1998;247:60-4.

[115] Rietjens S, Bast A, de Vente J, Haenen G. The olive oil antioxidant hydroxytyrosol efficiently protects against the oxidative stress induced impairment of the NO{middle dot} response of isolated rat aorta. *Am. J. Physiol. Heart Circ. Physiol.* 2006.

[116] Leger CL, Kadiri-Hassani N, Descomps B. Decreased superoxide anion production in cultured human promonocyte cells (THP-1) due to polyphenol mixtures from olive oil processing wastewaters. *J. Agric. Food Chem.* 2000;48:5061-7.

[117] Manna C, Galletti P, Cucciolla V, Moltedo O, Leone A, Zappia V. The protective effect of the olive oil polyphenol (3,4-dihydroxyphenyl)-ethanol counteracts reactive oxygen metabolite-induced cytotoxicity in Caco-2 cells. *J. Nutr.* 1997;127:286-92.

[118] Manna C, Migliardi V, Sannino F, De Martino A, Capasso R. Protective effects of synthetic hydroxytyrosol acetyl derivatives against oxidative stress in human cells. *J. Agric. Food Chem.* 2005;53:9602-7.

[119] Gutierrez VR, de la Puerta R, Catala A. The effect of tyrosol, hydroxytyrosol and oleuropein on the non-enzymatic lipid peroxidation of rat liver microsomes. *Mol. Cell Biochem.* 2001;217:35-41.

[120] de la Puerta R, Martinez Dominguez ME, Ruiz-Gutierrez V, Flavill JA, Hoult JR. Effects of virgin olive oil phenolics on scavenging of reactive nitrogen species and upon nitrergic neurotransmission. *Life Sci.* 2001;69:1213-22.

[121] Deiana M, Aruoma OI, Bianchi ML, Spencer JP, Kaur H, Halliwell B, et al. Inhibition of peroxynitrite dependent DNA base modification and tyrosine nitration by the extra virgin olive oil-derived antioxidant hydroxytyrosol. *Free Radic. Biol. Med.* 1999;26:762-9.

[122] Owen RW, Giacosa A, Hull WE, Haubner R, Spiegelhalder B, Bartsch H. The antioxidant/anticancer potential of phenolic compounds isolated from olive oil. *Eur. J. Cancer* 2000;36:1235-47.

[123] Manna C, Migliardi V, Golino P, Scognamiglio A, Galletti P, Chiariello M, et al. Oleuropein prevents oxidative myocardial injury induced by ischemia and reperfusion. *J. Nutr. Biochem.* 2004;15:461-6.

[124] Ochoa JJ, Huertas JR, Quiles JL, Olvera AB, Mataix J. Relative importance of the saponified and unsaponified fractions of dietary olive oil on mitochondrial lipid peroxidation in rabbit heart. *Nutr. Metab. Cardiovasc. Dis.* 1999;9:284-8.

[125] Visioli F, Caruso D, Galli C, Viappiani S, Galli G, Sala A. Olive oils rich in natural catecholic phenols decrease isoprostane excretion in humans. *Biochem. Biophys. Res. Commun.* 2000;278:797-9.

[126] Salvini S, Sera F, Caruso D, Giovannelli L, Visioli F, Saieva C, et al. Daily consumption of a high-phenol extra-virgin olive oil reduces oxidative DNA damage in postmenopausal women. *Br. J. Nutr.* 2006;95:742-51.

[127] Grasso S, Siracusa L, Spatafora C, Renis M, Tringali C. Hydroxytyrosol lipophilic analogues: enzymatic synthesis, radical scavenging activity and DNA oxidative damage protection. *Bioorg. Chem.* 2007;35:137-52.

[128] Trujillo M, Mateos R, Collantes de Teran L, Espartero JL, Cert R, Jover M, et al. Lipophilic hydroxytyrosyl esters. Antioxidant activity in lipid matrices and biological systems. *J. Agric. Food Chem.* 2006;54:3779-85.

[129] Galle J, Hansen-Hagge T, Wanner C, Seibold S. Impact of oxidized low density lipoprotein on vascular cells. *Atherosclerosis* 2006;185:219-26.

[130] Leenen R, Roodenburg AJ, Vissers MN, Schuurbiers JA, van Putte KP, Wiseman SA, et al. Supplementation of plasma with olive oil phenols and extracts: influence on LDL oxidation. *J. Agric. Food Chem.* 2002;50:1290-7.

[131] Grignaffini P, Roma P, Galli C, Catapano AL. Protection of low-density lipoprotein from oxidation by 3,4-dihydroxyphenylethanol. *Lancet* 1994;343:1296-7.

[132] Salami M, Galli C, De Angelis L, Visioli F. Formation of F2-isoprostanes in oxidized low density lipoprotein: inhibitory effect of hydroxytyrosol. *Pharmacol. Res.* 1995;31:275-9.

[133] Briante R, Febbraio F, Nucci R. Antioxidant/prooxidant effects of dietary non-flavonoid phenols on the Cu2+-induced oxidation of human low-density lipoprotein (LDL). *Chem. Biodivers.* 2004;1:1716-29.

[134] Turner R, Etienne N, Alonso MG, de Pascual-Teresa S, Minihane AM, Weinberg PD, et al. Antioxidant and anti-atherogenic activities of olive oil phenolics. *Int. J. Vitam. Nutr. Res.* 2005;75:61-70.

[135] Andrikopoulos NK, Kaliora AC, Assimopoulou AN, Papageorgiou VP. Inhibitory activity of minor polyphenolic and nonpolyphenolic constituents of olive oil against in vitro low-density lipoprotein oxidation. J. *Med. Food* 2002;5:1-7.

[136] Romani A, Lapucci C, Cantini C, Ieri F, Mulinacci N, Visioli F. Evolution of Minor Polar Compounds and Antioxidant Capacity during Storage of Bottled Extra Virgin Olive Oil. *J. Agric. Food Chem.* 2007.

[137] Wiseman SA, Mathot JN, de Fouw NJ, Tijburg LB. Dietary non-tocopherol antioxidants present in extra virgin olive oil increase the resistance of low density lipoproteins to oxidation in rabbits. *Atherosclerosis* 1996;120:15-23.

[138] Di Benedetto R, Vari R, Scazzocchio B, Filesi C, Santangelo C, Giovannini C, et al. Tyrosol, the major extra virgin olive oil compound, restored intracellular antioxidant defences in spite of its weak antioxidative effectiveness. *Nutr. Metab. Cardiovasc. Dis.* 2006.

[139] Haban P, Klvanova J, Zidekova E, Nagyova A. Dietary supplementation with olive oil leads to improved lipoprotein spectrum and lower n-6 PUFAs in elderly subjects. *Med. Sci. Monit.* 2004;10:PI49-54.

[140] Covas MI, Nyyssonen K, Poulsen HE, Kaikkonen J, Zunft HJ, Kiesewetter H, et al. The effect of polyphenols in olive oil on heart disease risk factors: a randomized trial. *Ann. Intern. Med.* 2006;145:333-41.

[141] Marrugat J, Covas MI, Fito M, Schroder H, Miro-Casas E, Gimeno E, et al. Effects of differing phenolic content in dietary olive oils on lipids and LDL oxidation--a randomized controlled trial. *Eur. J. Nutr.* 2004;43:140-7.

[142] Bellido C, Lopez-Miranda J, Perez-Martinez P, Paz E, Marin C, Gomez P, et al. The Mediterranean and CHO diets decrease VCAM-1 and E-selectin expression induced by modified low-density lipoprotein in HUVECs. *Nutr. Metab. Cardiovasc. Dis.* 2006;16:524-30.

[143] Lapointe A, Couillard C, Lemieux S. Effects of dietary factors on oxidation of low-density lipoprotein particles. *J. Nutr. Biochem.* 2006;17:645-58.

[144] Kaliora AC, Dedoussis GV, Schmidt H. Dietary antioxidants in preventing atherogenesis. *Atherosclerosis* 2006;187:1-17.

[145] Casalino E, Calzaretti G, Sblano C, Landriscina V, Felice Tecce M, Landriscina C. Antioxidant effect of hydroxytyrosol (DPE) and Mn2+ in liver of cadmium-intoxicated rats. *Comp. Biochem. Physiol. C Toxicol. Pharmacol.* 2002;133:625-32.

[146] Manna C, Galletti P, Cucciolla V, Montedoro G, Zappia V. Olive oil hydroxytyrosol protects human erythrocytes against oxidative damages. *J. Nutr. Biochem.* 1999;10:159-65.

[147] Quiles JL, Farquharson AJ, Simpson DK, Grant I, Wahle KW. Olive oil phenolics: effects on DNA oxidation and redox enzyme mRNA in prostate cells. *Br. J. Nutr.* 2002;88:225-34; discussion 223-4.

[148] Goya L, Mateos R, Bravo L. Effect of the olive oil phenol hydroxytyrosol on human hepatoma HepG2 cells : Protection against oxidative stress induced by tert-butylhydroperoxide. *Eur. J. Nutr.* 2007.

[149] Fki I, Sahnoun Z, Sayadi S. Hypocholesterolemic effects of phenolic extracts and purified hydroxytyrosol recovered from olive mill wastewater in rats fed a cholesterol-rich diet. *J. Agric. Food Chem.* 2007;55:624-31.

[150] Galletti P, Di Gennaro CI, Migliardi V, Indaco S, Della Ragione F, Manna C, et al. Diverse effects of natural antioxidants on cyclosporin cytotoxicity in rat renal tubular cells. *Nephrol. Dial. Transplant* 2005;20:1551-8.

[151] Kirkby K, Baylis C, Agarwal A, Croker B, Archer L, Adin C. Intravenous bilirubin provides incomplete protection against renal ischemia-reperfusion injury in vivo. *Am. J. Physiol. Renal. Physiol.* 2007;292:F888-94.

[152] Stupans I, Stretch G, Hayball P. Olive oil phenols inhibit human hepatic microsomal activity. *J. Nutr.* 2000;130:2367-70.

[153] Stupans I, Murray M, Kirlich A, Tuck KL, Hayball PJ. Inactivation of cytochrome P450 by the food-derived complex phenol oleuropein. *Food Chem. Toxicol.* 2001;39:1119-24.

[154] Yamazaki H, Inui Y, Wrighton SA, Guengerich FP, Shimada T. Procarcinogen activation by cytochrome P450 3A4 and 3A5 expressed in Escherichia coli and by human liver microsomes. *Carcinogenesis* 1995;16:2167-70.

[155] Aviram M, Rosenblat M. Paraoxonases and cardiovascular diseases: pharmacological and nutritional influences. *Curr. Opin. Lipidol.* 2005;16:393-9.

[156] Palmerini CA, Carlini E, Saccardi C, Servili M, Montedoro G, Arienti G. Activity of olive oil phenols on lymphomonocyte cytosolic calcium. *J. Nutr. Biochem.* 2005;16:109-13.

[157] Morris BJ. A forkhead in the road to longevity: the molecular basis of lifespan becomes clearer. *J. Hypertens* 2005;23:1285-309.

[158] Jeong J, Juhn K, Lee H, Kim SH, Min BH, Lee KM, et al. SIRT1 promotes DNA repair activity and deacetylation of Ku70. *Exp. Mol. Med.* 2007;39:8-13.

[159] Braam B, Langelaar-Makkinje M, Verkleij A, Bluyssen H, Verrips T, Koomans HA, et al. Anti-oxidant sensitivity of donor age-related gene expression in cultured fibroblasts. *Eur. J. Pharmacol.* 2006;542:154-61.

[160] Masella R, Di Benedetto R, Vari R, Filesi C, Giovannini C. Novel mechanisms of natural antioxidant compounds in biological systems: involvement of glutathione and glutathione-related enzymes. *J. Nutr. Biochem.* 2005;16:577-86.

[161] Masella R, Vari R, D'Archivio M, Di Benedetto R, Matarrese P, Malorni W, et al. Extra virgin olive oil biophenols inhibit cell-mediated oxidation of

LDL by increasing the mRNA transcription of glutathione-related enzymes. *J. Nutr.* 2004;134:785-91.

[162] Faine LA, Rodrigues HG, Galhardi CM, Ebaid GM, Diniz YS, Padovani CR, et al. Effects of olive oil and its minor constituents on serum lipids, oxidative stress, and energy metabolism in cardiac muscle. *Can. J. Physiol. Pharmacol.* 2006;84:239-45.

[163] Budiyanto A, Ahmed NU, Wu A, Bito T, Nikaido O, Osawa T, et al. Protective effect of topically applied olive oil against photocarcinogenesis following UVB exposure of mice. *Carcinogenesis* 2000;21:2085-90.

[164] Benavente-Garcia O, Castillo J, Lorente J, Alcaraz M. Radioprotective effects in vivo of phenolics extracted from Olea europaea L. leaves against X-ray-induced chromosomal damage: comparative study versus several flavonoids and sulfur-containing compounds. *J. Med. Food* 2002;5:125-35.

[165] D'Angelo S, Ingrosso D, Migliardi V, Sorrentino A, Donnarumma G, Baroni A, et al. Hydroxytyrosol, a natural antioxidant from olive oil, prevents protein damage induced by long-wave ultraviolet radiation in melanoma cells. *Free Radic. Biol. Med.* 2005;38:908-19.

[166] Nousis L, Doulias PT, Aligiannis N, Bazios D, Agalias A, Galaris D, et al. DNA protecting and genotoxic effects of olive oil related components in cells exposed to hydrogen peroxide. *Free Radic. Res.* 2005;39:787-95.

[167] Ragione FD, Cucciolla V, Borriello A, Pietra VD, Pontoni G, Racioppi L, et al. Hydroxytyrosol, a natural molecule occurring in olive oil, induces cytochrome c-dependent apoptosis. *Biochem. Biophys. Res. Commun.* 2000;278:733-9.

[168] Fabiani R, De Bartolomeo A, Rosignoli P, Servili M, Montedoro GF, Morozzi G. Cancer chemoprevention by hydroxytyrosol isolated from virgin olive oil through G1 cell cycle arrest and apoptosis. *Eur. J. Cancer Prev.* 2002;11:351-8.

[169] Della Ragione F, Cucciolla V, Criniti V, Indaco S, Borriello A, Zappia V. Antioxidants induce different phenotypes by a distinct modulation of signal transduction. *FEBS Lett.* 2002;532:289-94.

[170] Fabiani R, De Bartolomeo A, Rosignoli P, Servili M, Selvaggini R, Montedoro GF, et al. Virgin olive oil phenols inhibit proliferation of human promyelocytic leukemia cells (HL60) by inducing apoptosis and differentiation. *J. Nutr.* 2006;136:614-9.

[171] Guichard C, Pedruzzi E, Fay M, Marie JC, Braut-Boucher F, Daniel F, et al. Dihydroxyphenylethanol induces apoptosis by activating serine/threonine protein phosphatase PP2A and promotes the endoplasmic reticulum stress response in human colon carcinoma cells. *Carcinogenesis* 2006;27:1812-27.

[172] Nardini M, Leonardi F, Scaccini C, Virgili F. Modulation of ceramide-induced NF-kappaB binding activity and apoptotic response by caffeic acid in U937 cells: comparison with other antioxidants. *Free Radic. Biol. Med.* 2001;30:722-33.

[173] Martinez-Dominguez E, de la Puerta R, Ruiz-Gutierrez V. Protective effects upon experimental inflammation models of a polyphenol-supplemented virgin olive oil diet. *Inflamm. Res.* 2001;50:102-6.

[174] de la Puerta R, Martinez-Dominguez E, Ruiz-Gutierrez V. Effect of minor components of virgin olive oil on topical antiinflammatory assays. *Z. Naturforsch.* [C] 2000;55:814-9.

[175] Bitler CM, Viale TM, Damaj B, Crea R. Hydrolyzed olive vegetation water in mice has anti-inflammatory activity. *J. Nutr.* 2005;135:1475-9.

[176] Kohyama N, Nagata T, Fujimoto S, Sekiya K. Inhibition of arachidonate lipoxygenase activities by 2-(3,4-dihydroxyphenyl)ethanol, a phenolic compound from olives. *Biosci. Biotechnol. Biochem.* 1997;61:347-50.

[177] Petroni A, Blasevich M, Papini N, Salami M, Sala A, Galli C. Inhibition of leukocyte leukotriene B4 production by an olive oil-derived phenol identified by mass-spectrometry. *Thromb. Res.* 1997;87:315-22.

[178] de la Puerta R, Ruiz Gutierrez V, Hoult JR. Inhibition of leukocyte 5-lipoxygenase by phenolics from virgin olive oil. *Biochem. Pharmacol.* 1999;57:445-9.

[179] Moreno JJ. Effect of olive oil minor components on oxidative stress and arachidonic acid mobilization and metabolism by macrophages RAW 264.7. *Free Radic. Biol. Med.* 2003;35:1073-81.

[180] Beauchamp GK, Keast RS, Morel D, Lin J, Pika J, Han Q, et al. Phytochemistry: ibuprofen-like activity in extra-virgin olive oil. *Nature* 2005;437:45-6.

[181] Perez-Martinez P, Lopez-Miranda J, Blanco-Colio L, Bellido C, Jimenez Y, Moreno JA, et al. The chronic intake of a Mediterranean diet enriched in virgin olive oil, decreases nuclear transcription factor kappaB activation in peripheral blood mononuclear cells from healthy men. *Atherosclerosis* 2007.

[182] Maiuri MC, De Stefano D, Di Meglio P, Irace C, Savarese M, Sacchi R, et al. Hydroxytyrosol, a phenolic compound from virgin olive oil, prevents macrophage activation. *Naunyn. Schmiedebergs Arch. Pharmacol.* 2005;371:457-65.

[183] Visioli F, Bellosta S, Galli C. Oleuropein, the bitter principle of olives, enhances nitric oxide production by mouse macrophages. *Life Sci.* 1998;62:541-6.

[184] Dell'Agli M, Fagnani R, Mitro N, Scurati S, Masciadri M, Mussoni L, et al. Minor components of olive oil modulate proatherogenic adhesion molecules involved in endothelial activation. *J. Agric. Food Chem.* 2006;54:3259-64.

[185] Carluccio MA, Siculella L, Ancora MA, Massaro M, Scoditti E, Storelli C, et al. Olive oil and red wine antioxidant polyphenols inhibit endothelial activation: antiatherogenic properties of Mediterranean diet phytochemicals. *Arterioscler. Thromb. Vasc. Biol.* 2003;23:622-9.

[186] Ferrara LA, Raimondi AS, d'Episcopo L, Guida L, Dello Russo A, Marotta T. Olive oil and reduced need for antihypertensive medications. *Arch. Intern. Med.* 2000;160:837-42.

[187] Herrera MD, Perez-Guerrero C, Marhuenda E, Ruiz-Gutierrez V. Effects of dietary oleic-rich oils (virgin olive and high-oleic-acid sunflower) on vascular reactivity in Wistar-Kyoto and spontaneously hypertensive rats. *Br. J. Nutr.* 2001;86:349-57.

[188] Petroni A, Blasevich M, Salami M, Papini N, Montedoro GF, Galli C. Inhibition of platelet aggregation and eicosanoid production by phenolic components of olive oil. *Thromb. Res.* 1995;78:151-60.

[189] Leger CL, Carbonneau MA, Michel F, Mas E, Monnier L, Cristol JP, et al. A thromboxane effect of a hydroxytyrosol-rich olive oil wastewater extract in patients with uncomplicated type I diabetes. *Eur. J. Clin. Nutr.* 2005;59:727-30.

[190] Togna GI, Togna AR, Franconi M, Marra C, Guiso M. Olive oil isochromans inhibit human platelet reactivity. *J. Nutr.* 2003;133:2532-6.

[191] Calleja L, Paris MA, Paul A, Vilella E, Joven J, Jimenez A, et al. Low-cholesterol and high-fat diets reduce atherosclerotic lesion development in ApoE-knockout mice. *Arterioscler. Thromb. Vasc. Biol.* 1999;19:2368-75.

[192] Acin S, Navarro MA, Carnicer R, Arbones-Mainar JM, Guzman MA, Arnal C, et al. Dietary cholesterol suppresses the ability of olive oil to delay the development of atherosclerotic lesions in apolipoprotein E knockout mice. *Atherosclerosis* 2005;182:17-28.

[193] Acin S, Navarro MA, Arbones-Mainar JM, Guillen N, Sarria AJ, Carnicer R, et al. Hydroxytyrosol administration enhances atherosclerotic lesion development in apo E deficient mice. *J. Biochem.* (Tokyo) 2006;140:383-91.

[194] Norata GD, Marchesi P, Passamonti S, Pirillo A, Violi F, Catapano AL. Anti-inflammatory and anti-atherogenic effects of cathechin, caffeic acid and trans-resveratrol in apolipoprotein E deficient mice. *Atherosclerosis* 2006.

[195] Gonzalez-Santiago M, Martin-Bautista E, Carrero JJ, Fonolla J, Baro L, Bartolome MV, et al. One-month administration of hydroxytyrosol, a phenolic antioxidant present in olive oil, to hyperlipemic rabbits improves blood lipid profile, antioxidant status and reduces atherosclerosis development. *Atherosclerosis* 2006;188:35-42.

[196] Patrick L, Uzick M. Cardiovascular disease: C-reactive protein and the inflammatory disease paradigm: HMG-CoA reductase inhibitors, alpha-tocopherol, red yeast rice, and olive oil polyphenols. A review of the literature. *Altern. Med. Rev.* 2001;6:248-71.

[197] Visioli F, Bogani P, Grande S, Galli C. Mediterranean food and health: building human evidence. *J. Physiol. Pharmacol.* 2005;56 Suppl 1:37-49.

[198] Wahle KW, Caruso D, Ochoa JJ, Quiles JL. Olive oil and modulation of cell signaling in disease prevention. *Lipids* 2004;39:1223-31.

[199] Ursini F, Tubaro F, Rong J, Sevanian A. Optimization of nutrition: polyphenols and vascular protection. *Nutr. Rev.* 1999;57:241-9.

INDEX

D

E

F

G

H

I

J

K

L

M

N

O

P

Q

R

S

T

U

V

W

X

Y